Advances in Modeling Concrete Service Life

RILEM BOOKSERIES
Volume 3

RILEM, The International Union of Laboratories and Experts in Construction Materials, Systems and Structures, founded in 1947, is a non-governmental scientific association whose goal is to contribute to progress in the construction sciences, techniques and industries, essentially by means of the communication it fosters between research and practice. RILEM's focus is on construction materials and their use in building and civil engineering structures, covering all phases of the building process from manufacture to use and recycling of materials. More information on RILEM and its previous publications can be found on www.RILEM.net.

For further volumes:
http://www.springer.com/series/8781

Carmen Andrade • Joost Gulikers
Editors

Advances in Modeling Concrete Service Life

Proceedings of 4th International
RILEM PhD Workshop held in Madrid,
Spain, November 19, 2010

Editors
Carmen Andrade
Department of Phys. Chem.
of Bldg Materials
IETcc_CSIC
Serrano Galvache 4
28033 Madrid
Spain

Joost Gulikers
Centre for Infrastructure
Ministry of Infrastructure
and The Environment
Griffioenlaan 2
3526 LA Utrecht
Netherlands

ISSN 2211-0844 e-ISSN 2211-0852
ISBN 978-94-007-2702-1 e-ISBN 978-94-007-2703-8
DOI 10.1007/978-94-007-2703-8
Springer Dordrecht Heidelberg London New York

Library of Congress Control Number: 2011942225

© RILEM 2012
No part of this work may be reproduced, stored in a retrieval system, or transmitted in any form or by any means, electronic, mechanical, photocopying, microfilming, recording or otherwise, without written permission from the Publisher, with the exception of any material supplied specifically for the purpose of being entered and executed on a computer system, for exclusive use by the purchaser of the work.

Printed on acid-free paper

Springer is part of Springer Science+Business Media (www.springer.com)

Contents

Corrosion Data Interpretation in Concrete Structures 1
Carmen Andrade

**Pitfalls and Practical Implications in Durability
Design of Reinforced Concrete Structures** 11
Joost Gulikers

**Microstructure and durability of slag cement mortars
hardened under different relative humidity conditions** 21
José Marcos Ortega, Javier Sánchez, and Miguel Ángel Climent

**Study of corrosion rate variability
in indoor and outdoor specimens** 33
Filipe Pedrosa and Carmen Andrade

**From corrosion rate to accumulated corrosion
depth or loss in cross section of reinforcements** 43
Nuria Rebolledo and Carmen Andrade

**Steel Corrosion in a Chloride Contaminated Concrete
Pore Solution with Low Oxygen Availability** 53
Lina Toro, Carmen Andrade, José Fullea, Isabel Martínez,
and Nuria Rebolledo

**Application of risk analysis in structural
engineering – gas explosions** 65
Ramon Hingorani and Peter Tanner

**Hydrogen induced changes in structural
properties of iron: Ab initio calculations** 79
Alejandro Castedo, Javier Sanchez, José Fullea,
Carmen Andrade, and Pedro Luis de Andres

**Corrosion initiation and propagation
in cracked concrete – a literature review** 85
José Pacheco and Rob Polder

**Possibilities for improving corrosion protection of reinforced
concrete by modified hydrotalcites – a literature review** 95
Zhengxian Yang, Hartmut Fischer, and Rob Polder

**Determination of the probable failure mechanisms
and service life of offshore concrete gravity structures
in the OSPAR Maritime Area - research proposal** 107
Rod Jones, Moray Newlands, and Chris Thistlethwaite

**Electrically accelerated transport of chlorides
in concrete considering non-linear chloride binding
in non-equilibrium conditions**.. 119
Przemek Spiesz and Jos Brouwers

**Chloride binding related to hydration
products Part I: Ordinary Portland Cement**... 125
Miruna Marinescu and Jos Brouwers

Chloride ingress in cracked concrete- a literature review........................... 133
Branko Šavija and Erik Schlangen

**Numerical simulation of reinforcement corrosion
and protection in submerged hollow concrete structures**............................ 143
Andrea Della Pergola, Federica Lollini, Elena Redaelli,
and Luca Bertolini

RILEM Publications 153

RILEM Publications published by Springer 163

Corrosion Data Interpretation in Concrete Structures

Carmen Andrade

Abstract Concrete protects the steel embedded until carbonation front or the chloride threshold reaches the rebar. Then a corrosion process develops. In present paper the corrosion process is analyzed from its onset a definition of which is given. The techniques for its measurement are described and also are given a model for the corrosion propagation period as well as some alternatives to assume a value of the corrosion rate, even based in the values of the concrete resistivity. Finally, the functions of relation between the corrosion and the structural consequences are commented.

Keywords Concrete • Corrosion rate • Climate • Prediction

1 Introduction

The concrete is a good protector of the embedded steel until the carbonation front or the chloride threshold reach the rebar level. When the corrosion develops, four are the main structural consequences [1]: the decrease in the steel ductility, the steel cross section and the steel/concrete bond, and additionally, the stresses generated by the expansive character of the oxides produced lead into the cover cracking. Al theses consequences compromise the structural integrity and therefore it is crucial to design the concrete to sustain the environmental aggressivity and if the corrosion is produced during the structure life, to detect and prevent further attack. This is especially important in critical structures as those in nuclear installations. In present work it is described the corrosion initiation and propagation, what the corrosion rate generated means from a structural point of view, how to measure it on site and how to analyze the values from the standpoint of the calculation of the service life.

C. Andrade (✉)
Research Centre for Safety and Durability of Materials and Structures
CISDEM-UPM-CSIC, Madrid, Spain
e-mail: andrade@ietcc.csic.es

2 The Corrosion Process

2.1 Corrosion Onset

The corrosion develops when the chlorides, beyond a certain threshold level or the carbonation front reach the rebar. At that moment, identified as depassivation or corrosion onset, the passive layer dissolves and several iron oxides, described with the term "rust", start to form and grow.

The depassivation is in general localized, as the chlorides or the carbonation arrive first to the surface of the rebar closest to the external surface. When the corrosion progresses, the oxide formed, first diffuse through the pores and, after, induce the cracking of the concrete cover. The corroded area extends progressively in area and depth. The rate of this corrosion is defined as the amount of metal corroding referred to the unit of surface and time.

Having in mind the classical Service Life model [2] of considering an Initiation and a Propagation period, the first to briefly be considered is the fact that, the depassivation process cannot be understood as an instantaneous process. This is due to in the case of chlorides the threshold may vary from one part of the bar to other owing to the several unhomogeneities of the steel and the concrete. On the case of carbonation the front arrives first to the external part of the bar and therefore the corrosion increases as referred to the total surface. Then, events of activity-passivity may develop during a long period.

Therefore the reaching of a constant active corrosion state may last significantly. During this time however, the corrosion progresses and a localized loss in cross section is produced. The fact that the onset may last months or years introduces a certain difficulty on the definition of the end of the initiation period, t_i. However if the Accumulated corrosion, I_{ac} (Integration of the Corrosion Rate, I_{corr} with time) is plotted instead the instantaneous corrosion rate (Figure 1) the identification of the end of the t_i can be better defined by a certain loss in cross section, as indicated by the figure.

3 Measurement Techniques

The only electrochemical technique with quantitative ability regarding the corrosion rate is the so called Polarization Resistance, R_p [3]. This technique has been extensively used in the laboratory as in addition it gives the corrosion potential and the resistivity of the concrete [4–5]. It is based on the application of a small electrical perturbation to the metal by means of a counter and a reference electrode. Providing the electrical signal is uniformly distributed throughout the reinforcement, the $\Delta E/\Delta I$ ratio defines R_p. The corrosion current, I_{corr}, is inversely proportional to R_p, $I_{corr} = B/R_p$ where B is a constant. R_p can be measured by means of D.C. or A.C. techniques[5], both of which have specific features in order to obtain a reliable corrosion current value in agreement with gravimetric losses.

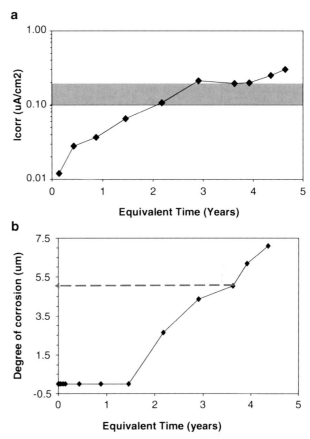

Fig. 1 a: Corrosion rate in function of the time. **b:** Accumulated corrosion (5 μm is suggested as limit of corrosion penetration P_x)

3.1 On site measurements

Direct estimation of True R_p values from $\Delta E/\Delta I$ measurements is usually unfeasible in large real concrete structures [6]. This is because the applied electric signal tends to vanish with distance from the counter electrode, CE rather than spread uniformly across the working electrode, WE. Therefore, the polarization by the electric signals not uniform and it reaches a certain distance that is named the critical length, L_{crit}. Hence, $\Delta E/\Delta I$ measurements on large structures using a small counter electrode provides an apparent polarization resistance ($R_{p,app}$) that differs from the true R_p value depending on the experimental conditions. Thus, if the metal is actively corroding, the current applied from a small CE located on the concrete surface is 'drained' very efficiently by the metal and it tends to confine itself on a small surface area. Conversely, if the metal is passive and R_p is high, the current applied tends to spread far away (e.g., around 50 cm) from the application point. Therefore, the apparent R_p approaches the true R_p for actively corroding reinforcement, but when the steel is passive, the large distance reached by the current needs a quantitative treatment.

Fig. 2 Modulated confinement of the current (guard ring) method

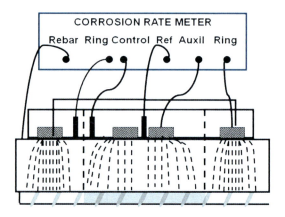

3.2 Modulated confinement of the current (guard ring) method [6]

There are several ways of accounting for a True R_p value, among which the most extended one is the use of a guard ring[6], in order to confine the current in a particular rebar area, as Figure 2 depicts. The measurement is made by applying a galvanostatic step, lasting 30-100 seconds, from the central counter. Then, another counter current is applied from the external ring, and this external current is modulated by means of the two reference electrodes called "ring controllers" in order to equilibrate internal and external currents, which enables a correct confinement, and therefore, calculation of R_p. By means of this electrical delimitation to a small zone of the polarized area, any localised spot or pit can be first, localised, and second its measurement can be made by minimising the inherent error of R_p. Not all guarded techniques are efficient. Only that using a "Modulated Confinement" controlled by two small sensors for the guard ring control placed between the central auxiliary electrode and the ring, shown in Figure 2, is able to efficiently confine the current within a predetermined area. The use of guard rings without this control leads into too high values of the I_{corr} for moderate and low values, and the error introduced in the case of localized pits, is very high.

3.3 Embedded sensors

The introduction of small sensors in the interior of the concrete, usually when placing it on-site is being one of the most promising developments in order to monitor the long term behaviour of the structures. The most usual, as in the case of non-permanent on-site techniques, is to embed reference electrodes or resistivity electrodes. They can inform of the presence of moisture and on the evolution of corrosion

Table 1 Ranges of the Corrosion rate and their relation to the velocity of the process

Range of V_{corr} (μm/year)	Corrosion progression
< 1	Negligible
1 – 5	Low
5 – 10	Moderate
> 10	High

potential. Others events that can be monitored are the advance of the carbonation or chloride fronts, the oxygen availability, temperature, concrete deformations and the corrosion rate.

A particular example of the use of embedded sensors is the case of storage facilities of low and medium radioactive wastes in El Cabril (Córdoba) [7]. There, a pilot container has been instrumented from 1995 by embedding 27 set of electrodes. The parameters controlled are: temperature, concrete deformation, corrosion potential, resistivity, and oxygen availability and corrosion rate. The impact of temperature on several of the parameters is remarkable, and therefore, care has to be taken when interpreting on-site results.

3.4 Ranges of corrosion rate values measured on-site

The experience from real structures has confirmed the ranges of corrosion current values previously recorded in laboratory experiments. In general, values of corrosion rates higher than $1 \mu A/cm^2$ are seldom measured while values between 0.1-$1 \mu A/cm^2$ are the most frequent. When the steel is passive very low values (smaller than 0.05-$0.1 \mu A/cm^2$) are measured. In Table 1 are given the ranges of V_{corr} and their relation to how fast is the corrosion process progressing.

4 Corrosion Evolution

4.1 Influence of environment

The two parameters that influence most the corrosion rate are the moisture content in the concrete pores and the temperature. The moisture depends on the atmospheric Relative Humidity and on the raining intensity [8]. Then, the levels of corrosion rate and its changes during time will depend on the climatic parameters.

One can expect that if the concrete is maintained in chambers with constant humidity and temperature, the corrosion rate would remain constant as well. However, this is not happening because the corrosion is a dynamic process in which the formation of oxides is continuous as well as the extension or increase in depth of the local attack, which make the circumstances different each new time step.

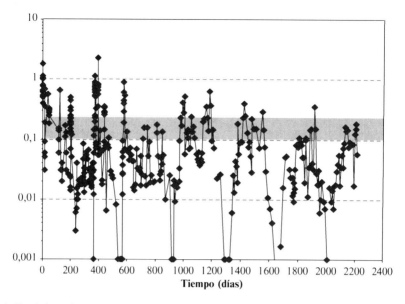

Fig. 3 Evolution of I_{corr} of the reinforcement of a concrete specimen fully carbonated and placed unsheltered from rain in Madrid atmosphere

Thus, Figure 3 shows the corrosion rate evolution of a concrete beam having in the mixing water 3% $CaCl_2$ by cement weight and exposed to Madrid atmosphere non-sheltered from rain. The evolution of corrosion rate in open atmosphere is shown in Figure 3 where the changes of the I_{corr} with time are plotted showing an apparent scatter of values difficult of being interpreted. However, when plotted as Accumulated corrosion, I_{ac}, (Figure 4) the trend is much clearer and can be used for the calculation of the loss in cross section or Corrosion Penetration P_{corr}. The slope of the plot of the I_{ac} gives the Representative corrosion rate, $V_{corr, REP}$ which results almost constant.

4.2 Determination of a Representative Corrosion Rate

A "Representative corrosion rate, $V_{corr, REP}$" can be established in different manners which are summarized in Table 2. In existing structures where the corrosion process develops one manner is, if electrochemical measurements are taken, from the slope of the integration of the instantaneous I_{corr}-time curves. Also in existing structures when no electrochemical techniques are available, it is possible to obtain and averaged value from measuring the loss in diameter of the bars and dividing by the number of years of corrosion propagation.

However in new structures or in those which are not yet corroding, another approach to estimate annual averaged values for $V_{corr, REP}$ consists in ascribing a value in function of the aggressivity of the particular environment (for instance the exposure classes of EN 206, see Table 3) or from other parameters, such as the evaporable water content or the values of concrete resistivity [3, 7, 9].

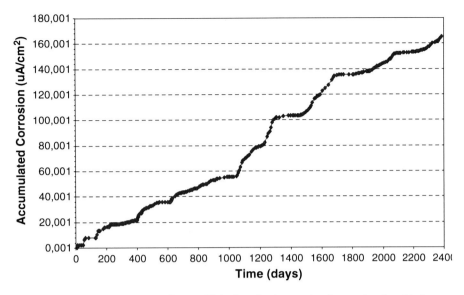

Fig. 4 Integration of the previous figure which gives the Accumulated corrosion, I_{ac} with time

Table 2 Possibilities of estimation of the averaged corrosion rate

Available possibilities of estimation	Method of estimation of $V_{CORR, REP}$
1) From on site electrochemical measurements of I_{corr}	***Integration*** of the curves I_{corr}- time and plotting of the slope
2) From de measuring of the diameter loss	Dividing by the number of years of corrosion propagation
3) From exposure classes	By ascribing a value (mean or characteristic) to each ***ambient class***. See Table 3
4) Indirectly, from a ***concrete characteristic*** related to I_{corr} as pore water content or concrete Resistivity	a) From ***pore water content*** by an algorithm b) From concrete ***Resistivity*** taking account of its evolution with time and with pore water content

5 Engineering Application of the Corrosion Rate

5.1 Basic Model of Corrosion Propagation

A model for the propagation Period has to express the advance of the corrosion with time [10]. As the accumulated corrosion has been defined as corrosion penetration P_{corr} or loss in diameter. The model can be based in the multiplication of the corrosion rate V_{corr} multiplied by the time:

$$P_{corr}(mm) = V_{corr}(mm/year) \cdot t\,(years)$$

Table 3 Averaged corrosion rates $V_{corr, REP}$ for the exposure classes of EN206

Exposure class		$V_{corr, REP}$ (μm/year)
0	No risk of corrosion, very dry	0
XC1	Dry or permanent wet	0
XC2	Wet rarely dry	4
XC3	Moderate humidity	2
XC4	Cyclic wet dry	5
XD1	Moderate humidity	4
XD2	Wet, rarely dry	30
XD3	Cyclic, wet and dry	30
XS1	Airborne salt conditions	30
XS2	Submerged	Not corrosion expected or 10
XS3	Tidal, splash and spray zones	70

5.2 Damage functions for the Structural consequences of the corrosion

The corrosion of the reinforcement aims into several important structural damages [11–13]. The main are known to be:

- The loss of bar diameter
- The loss of metal ductility
- The cracking of concrete cover with the consequential loss in concrete cross section
- The loss in reinforcement/concrete bond

Corrosion can be treated as a "limit state" in which several effects or consequences should be verified. Figure 5 [11] shows an illustration of the several effects or limit states due to corrosion to be verified in function of P_x.

The verification implies the calculation of the time to reach a certain limit state, which can be now expressed in terms of corrosion propagation, t_p, as:

$$t_p = \frac{P_{lim}}{V_{corr,REP}},$$

where: P_{lim} is the chosen limit of the attack penetration P_{corr} (mm) for the Limit state considered. Then, the reduced section and the loss in bond and the cover cracking should be taken into consideration for the recalculation of the damaged element. Table 4 gives some "damage functions" [12–16] relating the consequences of corrosion with the advance of the process or corrosion penetration P_{corr}.

Fig. 5 Possible limit states due to corrosion to be verified in function of corrosion penetration

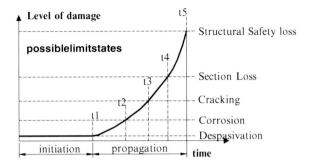

Table 4 Some damage functions relating consequences of corrosion and corrosion penetration

Consequences of corrosion	Damage function
loss of bar diameter	P_{corr} (mm) = $0.0116 \cdot I_{corr, mean} \cdot \alpha \cdot t$
loss of metal ductility	Damage function not produced yet
Cover (C in mm) cracking (w = crack width in mm, $\emptyset$ = bar diameter in mm and f_{ct} = indirect splitting stregth)	$w = k \left(\dfrac{P_x}{C/\phi} \right)$ being k = 5-10 $P_{X0} = 83.8 \cdot 10^{-3} + 7.4 \cdot 10^{-3} \left(\dfrac{C}{\Phi} \right) - 22.6 \cdot 10^{-3} f_{ct,sp}$
loss in reinforcement/ concrete bond (f_b)	With stirrups : $f_{b=}$ 5.25 - 2.72 P_{corr} With external pressure p: $f_b = (4.75 - 4.64\ P_x)/(1 - 0.098\ p)$

6 Final Comments

In the design of a structure is necessary to take into account the environment and its aggressivity in order to identify the degradation processes that may occur and to provide the measures to avoid the reinforcement corrosion during the service life of the structure. Although still accurate models of service life do not exist, as the available ones have not been calibrated, some approximate calculations can be undertaken. In them is necessary to identify the sensitive zones of the structure and to calculate the time to the corrosion onset and which will the progression of the corrosion if developed. It is desirable to foresee in these sensitive zones the monitoring of the possible appearance of reinforcement corrosion. Once the loss in steel cross section is calculated, the rest of structural consequences can be deduced from the modification of present formulas in the Codes taken into account the reduced properties of the concrete and the reduced sections and steel/concrete bond.

Acknowledgements The authors would like to acknowledge the financing of the Ministry of Science and Innovation for the INGENIO 2010-CONSOLIDER Project on "Safety and Durability of Structures: SEDUREC".

References

[1] Andrade, C., Alonso, C. and Molina, F. J.: "Cover cracking as a function of rebar corrosion: Part I – Experimental test". Materials and Structures, 26, pp. 453–464, 1993.
[2] Tuutti, K. – Corrosion of steel in concrete", Swedish Cement and Concrete Institute, (CBI) no. 4-82, Stockholm (1982) 486
[3] Andrade, C. and Gónzalez, J.A., "Quantitative measurements of corrosion rate of reinforcing steels embedded in concrete using polarization resistance measurements", *Werkst. Korros.*, **29**, 515 (1978).
[4] ASTM C876-91. "Standard Test Method for Half Cell Potentials of Uncoated Reinforcing Steel n Concrete".
[5] Elsener, B and Bóhni, H. Corrosion Rates of Steel in Concrete, N.S. Berke, V.Chaker and D. Whiting (Eds.), *ASTM STP 1065*, pp. 143–156. (1990).
[6] Feliú, S., González, J.A., Feliú, S.Jr., and Andrade, C., "Confinement of the electrical signal or in-situ measurement of Polarization Resistance in Reinforced concrete," *ACI Mater. J.*, **87**, pp 457. (1990).
[7] Andrade C; Sagrera J.L; Gonzalez J.A; Jiménez F; Bolaño J.A; Zuloaga P. "Corrosion monitoring of concrete structures by means of permanent embedded sensors". *Niza. Eurocorr'96*.
[8] C. Andrade, A. Castillo "Evolution of reinforcement corrosion due to climatic variations". Materials and Corrosion. Vol 54, Issue 6, pags 379–386
[9] Millard, S.G. and Gowers, K.R., "Resistivity assessment of in-situ concrete: the influence of conductive and resistive surface layers", *Proc. Inst. Civil Engrs. Struct. & Bldgs*, **94**, paper 9876, pp.389–396. (1992).
[10] Andrade, C., Alonso, C., Rodríguez, J., "Remaining service life of corroding structures", IABSE Symposium on Durability, Lisboa, (Sep 1989): pp. 359–363.
[11] DURACRETE. Probabilistic performance based on durability design of concrete structures. EU-Brite EuRam Project BE95–1347. A number of reports available from CUR Centre for Civil Engineering Research and Codes, Gouda, The Netherlands.
[12] Rodríguez, J.; Ortega, L.M.; Casal, J. y Díez, J.M : *Corrosion of reinforcement and service life of concrete structures*, Proceedings of Durability of Building Materials and Components Stockholm (1996) 117–126, Vol.I, Edited by C. Sjöstrom, E&FN Spoon Publishers
[13] Torres Acosta, A.A.: "Cracking induced by localized corrosion of reinforcement in chloride contaminated concrete". Ph. D. Thesis, University of South Florida, Florida, USA, 1999.
[14] Vidal, T., Castel, A. and Francois, R.: "Analyzing crack width to predict corrosion in reinforced concrete", Cement and Concrete Research 34, 2004, pp. 165–174.
[15] Martín-Perez, B.: "Service life modeling of RC highway structures exposed to chlorides". Ph.D. dissertation, Dept. of Civil Engineering, University of Toronto, 1998.
[16] CONTECVET IN30902I, "A validated users manual for assessing the residual life of concrete structures", DG Enterprise, CEC, (2001). (The manual for assessing reinforced structures affected by reinforcement corrosion can be seen at the web sites of IETcc (www.ietcc.csic.es) and GEOCISA (www.geocisa.es))

Pitfalls and Practical Implications in Durability Design of Reinforced Concrete Structures

Joost Gulikers

Abstract Design for durability of infrastructure facilities is becoming increasingly important in view of the large economical impact of premature maintenance and repair. Internationally there is a trend to shift from a prescriptive to a performance-based approach. In scientific literature most attention is devoted to the development of performance criteria for durability by using semi-probabilistic and probabilistic approaches. However, the practical implications, the eventual benefits, if any, the limitations and pitfalls of the mathematical models underlying a performance-based approach are often not addressed. With respect to resistance to chloride ingress a performance-based design approach requires the availability of a simple, cheap and fast test method to quantify the potential performance of a concrete composition. In this paper the chloride migration test being the most commonly used test method will be critically analysed. In addition, experience in practice obtained for "real-crete" has demonstrated a very wide scatter over time of the migration coefficient even for a fixed concrete composition. This experience combined with the serious questions on the overall benefits will make owners, contractors and concrete producers hesitant to implement a performance-based approach for durability design.

1 Introduction

With respect to design of reinforced concrete structures a clear trend can be observed worldwide to change from prescription specifications to performance specifications with most of the interest being focused on performance for durability. Several objective arguments can be raised to support a performance-based approach, the most logical one being the fact that physical properties of hardened concrete are of importance

---

J. Gulikers (✉)
Ministry of Infrastructure and the Environment,
Rijkswaterstaat- Centre for Infrastructure, Utrecht, Netherlands
e-mail: joost.gulikers@rws.nl

rather than concrete composition in terms of water-cementitious ratio, and cement and air content. In addition, the present codes do not make any distinction between types of cement although it is well known that the use of supplementary cementitious materials, such as fly ash, granulated ground blast furnace slag and silica fume, as additions or in blended cements, have proven to result in better durability performance of concrete structures.

Generally, chloride-induced reinforcement corrosion is considered the major threat to the long-term durability of infrastructure facilities impairing both serviceability (staining, cracking and spalling of the concrete cover) and structural integrity (loss of steel cross sectional area, loss of bond between steel and concrete). Consequently, most of the attention for developing a performance-based approach in durability design is directed at chloride ingress. The most commonly heard argument in scientific articles to support a change from a prescriptive to a performance-based approach is that in a considerable number of cases chloride-induced reinforcement corrosion is suggested to have caused significant premature damage resulting in serious maintenance activities. However, in most Dutch infrastructure facilities premature corrosion-induced damage is of a limited extent, even after more than 50 years of service. In the relatively limited number of situations where premature damage had occurred, it was established that in the majority of cases a low cover depth, poor compaction and inadequate curing were the main causes. Thus the conclusion is justified that the principal cause for the occurrence of premature damage is simply not meeting the requirements as given in the codes. Consequently, it has to be appreciated that a performance-based approach will neither provide a proper solution to the assumed problem as long as adequate quality control on site cannot be assured.

In order to allow for the application in practice of a performance-based approach widely accepted criteria should be available as well as reliable, consistent and standardized test procedures to quantify the relevant durability properties [1]. In general, to be accepted by the major parties involved, i.e. the owner, contractor and concrete producer, a performance-based approach should result in an improvement of the current deemed-to-satisfy approach. In addition, such an approach should be based on realistic requirements, be manageable and eventually be cost-effective.

2 Mathematical Backgrounds for the Derivation of Performance Criteria

With respect to chloride-induced corrosion the combination of quantity, i.e. thickness, and quality, i.e. permeability, of the concrete cover determines the structure's resistance to chloride ingress. Most often the quality of the concrete is expressed in terms of a diffusion coefficient.

In order to derive a quantifiable relationship between thickness and diffusivity a mathematical model is needed. The most commonly applied engineering model to

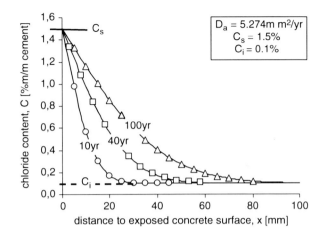

Fig. 1 Theoretical chloride profiles calculated according to Eq. (1) using a constant D_a

describe chloride ingress into concrete is based on Fick's second law of diffusion, generally referred to as the error function solution, as presented by (see Figure 1):

$$C(x,t) = C_s - (C_s - C_i) \cdot erf\left(\frac{x}{2\sqrt{D_a \cdot t}}\right) \quad (1)$$

with

$C(x,t)$	chloride content at depth x and time t, [%m/m cement];
C_s	chloride content at exposed concrete surface, [%m/m cement];
C_i	initial chloride content resulting from contamination of the concrete mix ingredients, [%m/m cement];
x	distance to the concrete surface exposed to chlorides, [mm];
D_a	apparent chloride diffusion coefficient, [mm²/yr];
erf	error function.

In addition, an acceptance criterion should be established. With respect to chloride ingress it is generally agreed upon that for new concrete structures end of service life is reached when depassivation of the embedded steel reinforcement occurs. This implies that the design service life, t_{dsl}, will be reached when the chloride content at the level of the reinforcing steel, i.e. for $x = c$, attains the so-called critical chloride content, C_{crit}. By re-arranging Eq. (1) this results into a mathematical relationship between thickness of concrete cover, c, and concrete quality, as expressed by the apparent diffusion coefficient, D_a, according to:

$$D_a = \frac{c^2}{A \cdot t_{dsl}} \quad (2)$$

with the factor A defined by:

$$A = 2 \cdot \text{inverf}\left(\frac{C_s - C_{crit}}{C_s - C_i}\right) \quad (3)$$

As for a given concrete composition and exposure conditions the magnitudes of C_s, C_i and C_{crit} may be considered to be constant, this implies that the factor A also represents a constant.

The apparent chloride diffusion coefficient D_a, has been found to be time-dependent according to the empirical relationship (power law):

$$D_a(t) = D_a(t_{ref}) \cdot \left(\frac{t_{ref}}{t}\right)^a \quad (4)$$

with

$D_a(t_{ref})$ apparent diffusion coefficient at a reference (concrete) age, [mm²/yr];
t_{ref} reference concrete age, [yr];
a ageing exponent, $0.0 \leq a \leq 1.0$ [-].

Since for new concrete structures the apparent diffusion coefficient at $t = t_{dsl}$ is unknown, its value is often predicted by the use of results obtained from an accelerated laboratory test performed on specimens made of the concrete composition to be used, possibly combined with a number of correction factors. In DuraCrete [2] the rapid chloride migration [3] test has been adopted as the accepted test method to quantify the potential concrete quality with respect to resistance against chloride ingress. This test result is multiplied by correction factors for environmental and curing conditions, referred to by k_e and k_c respectively, in order to arrive at practical levels of the diffusion coefficient according to:

$$D_a(t_{ref}) = k_e \cdot k_c \cdot D_{nssm}(t_{ref}) \quad (5)$$

with

D_{nssm} chloride migration coefficient obtained from an accelerated non-steady state laboratory experiment at age t_{ref}, [mm²/yr].

According to [2] the environmental factor k_e is dependent not only on the humidity conditions but also on the type of cement.

Eventually, this results in a modified relationship between thickness of concrete cover, c, and concrete quality, characterized by D_{nssm}, according to:

$$D_{nssm} = \left(\frac{c}{2A}\right)^2 \cdot \frac{1}{k_e \cdot k_c \cdot \left(\frac{t_{ref}}{t_{dsl}}\right)^a \cdot t_{dsl}} \quad (6)$$

Pitfalls and Practical Implications in Durability Design 15

Generally, the migration test is performed at a concrete age of 28d and consequently t_{ref} = 28d has to be introduced into the equations. It has to be noted that the correction factors k_e and k_c have been calibrated for this reference age. However, it is often overlooked by users of this so-called factorial approach that in DuraCrete the quantification of the ageing exponent, a, as well as the correction factors k_e and k_c, for most types of cement is based on expert opinion due to the limited availability of long term performance indicators.

3 Pitfalls of Relying on Mathematical Models

The use of an unambiguous criterion is an essential requirement for a performance-based approach. As all model parameters in Eq. (6) are of a stochastic nature a probabilistic approach is often advocated for durability design, necessitating that the criterion is expressed in terms of an acceptable probability of depassivation, P_{dep}. Apparently a probability level $P_{dep} \approx 0.50$, resulting from a deterministic calculation employing mean values for all model parameters, is considered too high as such a level is assumed to result in a significant amount of damage (cracking and spalling). As an example, in the Netherlands P_{dep} = 0.10 [4] is considered acceptable whereas in Germany P_{dep} = 0.30 is preferred [5]. However, it has to be concluded that the actual physical meaning of P_{dep} still remains largely unclear.

In order to prevent the application of time consuming full probabilistic calculations, in the Dutch CUR Guideline 1 [4] a pseudo-probabilistic approach has been advocated by introducing an allowance for concrete cover thickness Δc = 20mm for reinforcing steel and Δc = 30mm for prestressing steel as to compensate for the use of mean values for all model parameters. It is argued that for reinforcing steel this procedure will result in a probability of depassivation P_{dep} being "less than approximately" 10% and for prestressing steel P_{dep} is considered to be "less than approximately" 5%.

Introducing a cover allowance, Δc, a modified relationship between concrete quality, as characterized by the mean migration coefficient μD_{nssm}, and mean thickness of concrete cover, μc, can be established:

$$\mu D_{nssm}(28d) = \left[\frac{(\mu c - \Delta c)}{2A}\right]^2 \cdot \frac{1}{k_e \cdot k_c \cdot \left(\dfrac{0.767}{t_{dsl}}\right)^a \cdot t_{dsl}} \tag{7}$$

CUR Guideline 1 provides tables for t_{dsl} = 80, 100, and even 200 years in which combinations of μc and the maximum value of μD_{nssm}(28d) are presented. Figure 2 shows the results for t_{dsl} = 100yr for concrete components exposed to a de-icing salt environment referred to by XD, i.e. the exposure classes XD1, XD2 and XD3 as defined according to [6]. The significant difference between concrete made with CEM II/B-V (Portland cement with fly ash) and concrete made with other types of cement is mainly due to the extremely high level of the ageing exponent μa = 0.80 being used (CEM I: μa = 0.40; CEM III/A: μa = 0.65; CEM III/B: μa = 0.70). However, in [7] it

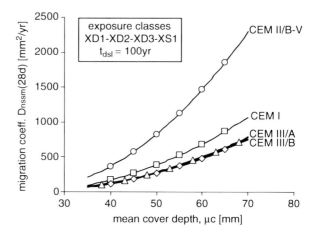

Fig. 2 Relationship between cover depth, c, and migration coefficient, $D_{nssm}(28d)$ according to [4] for $t_{dsl} = 100yr$

has been clearly demonstrated that such a pseudo-probabilistic approach based on a constant cover allowance Δc is deemed to fail as for reinforcing steel it will result in a wide range of probabilities. For the input used in [4] P_{dep} will range from 0.05 to 0.22, the latter being significantly in excess of the aimed level of $P_{dep} \approx 0.10$. Thus in fact, CUR Guideline 1 allows 2 sets of contradicting rules of play to be used and it is anticipated that for the owner this will result in major problems in construction practice.

Another essential requirement for the derivation of performance criteria based on mathematical models is that this has to result in a realistic output, e.g. for concrete cover thickness. In most mathematical models implicit assumptions are incorporated which may seriously limit the field of application. Eq. (4) predicts a gradual decrease of the apparent chloride diffusion coefficient, D_a, over time as dictated by the power law. This gradual decrease is said to be due to ongoing cement hydration and drying out of the concrete cover zone. However, hydration will effectively stop within 20 to 50 years or even at an earlier age, dependent o the type of cement, and drying out will not continue until infinity as within an exposure period of 50 years it is likely that a dynamic equilibrium with the environment (relative humidity, wetting and drying cycles) will be reached. Unfortunately, most designers but also academics are insufficiently aware of these limitations and implicit assumptions of mathematical models and thus frequently durability designs are made without any check of the veracity of both the input as well as the output. As an example, Figure 3 shows the relationship between design service life, t_{dsl}, and mean cover depth, μc, according to the calculation procedure and input values of CUR Guideline 1 [4], for concretes made with 4 different types of cement and exposed to a de-icing salt environment (exposure classes XD1, XD2, and XD3).

According to [4], for CEM I an increase of the cover depth of 6.8mm is needed whereas for CEM II/B-V an extra thickness of 2.6mm will suffice in order to increase the design service life from 50 to 200 years.

For an owner these calculated results seem hardly realistic and will thus merely lead to a disqualification of the consultant advocating such cover depths. Moreover, concrete structures with a design service life in excess of 100 years will undoubtedly be of particular importance and consequently special attention will be paid to durability design rather than by simply adopting a table.

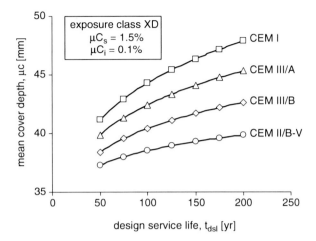

Fig. 3 Relationship between t_{dsl} and μc according to [4] for concretes made with CEM I, CEM III/A, CEM III/B and CEM II/B-V characterized by $D_{nssm}(28d)$ = 252.5, 157.8, 126.2 and 315.6 mm²/yr, respectively, for exposure class XD

In addition it has to be concluded that in most situations a full probabilistic approach is not feasible as most model parameters cannot be statistically quantified at a reliable level. Consequently "expert" guessing is often used although this is generally not explicitly mentioned in the documents. Moreover, a probabilistic approach is prone to manipulation of input data and is often misused merely as a selling argument to convince owners and asset managers.

4 Accelerated Chloride Migration Test

In view of the need to provide results on concrete quality in a short period of time and at low cost, the so-called rapid chloride migration test as performed according to [3] has frequently been advocated. The application of an electrical voltage of approximately 30V over a concrete disc of 50mm height, corresponding to an electrical field of 600 V/m, will result in a significant acceleration of chloride ingress within 24 hours. The resulting chloride penetration depth is measured on a freshly split sample using a silver nitrate solution sprayed on the concrete surface resulting in coloration. The average penetration depth is then translated into a migration coefficient, D_{nssm}, through a mathematical expression.

Theoretically, such a pronounced electrical field gives rise to the development of a tsunami-shaped chloride profile, which is distinctly different from the chloride profile resulting from a long-term bulk-diffusion test, see Figure 4. Accordingly, the chloride concentration will drop within a distance of a few millimetres and thus a distinct chloride penetration front will result. It is clear that such a steep profile is advantageous with respect to determination of the penetration depth as it will largely be independent of the actual chloride concentration at which coloration occurs. However, the underlying theory has come under discussion as the actually measured chloride profiles clearly demonstrate that the predicted tsunami profile does not develop, see Figure 5 [8]. On the contrary, these measured chloride profiles seem to be more comparable to a traditional diffusion profile. In view of the widespread use

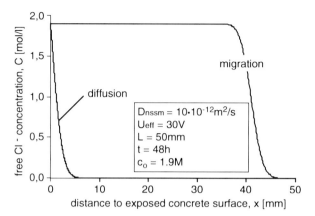

Fig. 4 Calculated theoretical chloride profiles resulting from ideal migration and ideal diffusion, both after 48h of testing

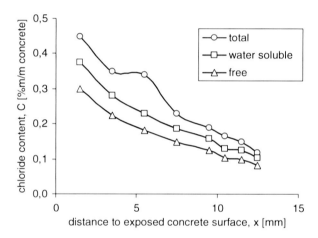

Fig. 5 Measured chloride profile resulting from an accelerated migration test [7]

of this test in practice further in-depth research is urgently required to reveal the actual transport mechanisms to validate the measured chloride profiles.

5   Realcrete Versus Labcrete

Generally, D_{nssm} is determined on concrete samples obtained from specimens mixed, placed, compacted, cured and stored under optimal laboratory conditions, commonly referred to as 'labcrete'. Moreover, the samples for migration testing are retrieved from the core part of the specimens, mostly cubes 150mm. Consequently, it becomes questionable to which extent the concrete quality determined on this labcrete reflects the actual quality of the 'covercrete' achieved on site. An impression on the difference between labcrete and realcrete is demonstrated in Figure 6. A concrete producer requested a laboratory to frequently perform migration tests on

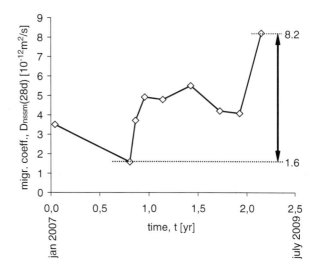

Fig. 6 Migration coefficients D_{nssm}(28d) measured on realcrete with a fixed concrete composition during a period of 2.5 yr

concrete samples based on a standard mix design during a period of more than 2 years. The concrete specimens used were obtained from realcrete produced in large quantities (9m³), before leaving the production plant. The results on D_{nssm} show a very wide scatter with values ranging from 1.6 and 8.2 10^{-12}m²/s, corresponding to 50.5 and 258.8 mm²/yr, respectively. In terms of thickness of concrete cover these extremes would imply a range from 30 to 60mm. Despite the fact that systematic investigations were performed, the concrete producer was not able to find the major causes of this variation. Apparently, this scatter has to be regarded as a common natural variation. These findings have been confirmed by other Dutch concrete producers. Although this example could be regarded as quite extreme, it highlights the significant difference between labcrete and realcrete.

In addition it has to be borne in mind that in the situation described here the concrete produced had not yet been processed by the contractor on site, i.e. effects of placement, compaction and curing are not included in these results.

6 Concluding Remarks

With respect to durability design of reinforced concrete structures a trend can be observed to change from prescriptive to performance-based requirements. To a considerable extent this trend is driven by scientists advocating the use of mathematical models for service life design introducing either a semi-probabilistic or a full probabilistic approach. Often too much emphasis is placed on the mathematical model rather than on the eventual benefits of a performance-based approach for the major stakeholders, i.e. the owner, contractor and concrete producer.

Users of mathematical degradation models are insufficiently aware of its limitations and implicit assumptions. Moreover, most of the input is eventually based on expert opinion, engineering judgement, or best guesses, rather than on data obtained from concrete structures using an objective and sound interpretation. Thus it is not surprising to note that design for durability of reinforced concrete structures turns out to become a number's game as to arrive at the desired output. In a number of cases the poor understanding of the limitations of mathematical models has resulted in unrealistic predictions for thickness of concrete cover.

The practical aspects of a performance-based approach seem to be forgotten as most attention is directed at advanced modelling. However, the modeller should be aware that input values derived from accelerated laboratory tests merely reflects the concrete quality achieved under ideal production and curing conditions and practice has clearly demonstrated that these results are hardly representative for the quality of realcrete as achieved in the cover zone. Moreover, research on the rapid chloride migration test has indicated that the theory underlying the rapid chloride migration test is not sound. Consequently, doubts are raised on the results and the relationship of the chloride migration coefficient with practice.

It has to be appreciated that adequate and strict quality control on site (still) remains the most effective and efficient way to achieve durable concrete structures.

References

[1] Bickley, J., Hooton, R.D. and Hover, K.C. (2006), 'Preparation of a performance-based specification for cast-in-place concrete', RMC Research Foundation.
[2] DuraCrete (2000), 'Final Technical Report', Doc. BE95-1347/R17, CUR, Gouda.
[3] NT Build 492 (1999), 'Chloride migration coefficient from non-steady-state migration experiments', nordtest, Espoo.
[4] CUR Guideline 1 (2009), 'Durability of structural concrete with respect to chloride-induced reinforcement corrosion', CUR, Gouda (in Dutch).
[5] Gehlen, J., Schiessl, P. and Schiessl-Pecka, A (2008) *Beton- und Stahlbetonbau*, vol. **103**, n. 12, p. 840 (in German).
[6] EN 206-1 (2001), 'Concrete Part 1: Specification, performance, production and conformity', CEN, Brussels.
[7] Gulikers, J. (2010), in *Service Life Design for Infrastructure*, Proceedings of an International Conference, pp. 359-368, RILEM, PRO70.
[8] Yuan, Q. (2009), 'Fundamental studies on test methods for the transport of chloride ions in cementitious materials', PhD Thesis, University of Ghent, Ghent.

Microstructure and durability of slag cement mortars hardened under different relative humidity conditions

José Marcos Ortega, Javier Sánchez, and Miguel Ángel Climent

Abstract In recent years the use of active additions on cement, like ground granulated blast-furnace slag, has become very popular because they have many environmental benefits. In this work, the behaviour of mortars made with two different cement types, an ordinary Portland cement and a slag cement, was tested. These mortars were exposed to two different environmental conditions during their hardening, an optimum laboratory condition and a Mediterranean climate environment with lower relative humidity. The development of the microstructure of mortars and the changes of their service properties were studied at different hardening ages until 90 days. The mortars made with slag cement have shown a good behaviour compared with ordinary Portland cement mortars. The relative humidity has an influence on properties of both cements studied. As a preliminary conclusion, cements with slag hardened under environmental conditions of a Mediterranean climate, can have good service properties at early hardening ages, even better than Portland cement.

1 Introduction

The use of mineral admixtures, like fly ash or ground granulated blast-furnace slag, in the cement manufacture has many environmental benefits. The main advantages are the reduction of CO_2 emissions and the lower energy consumption during the cement production. The particular case of ground granulated blast-furnace slag, and their effect on the properties of the cementitious materials is a topic of study. Many studies show that in laboratory conditions this kind of material has good service properties, even better than Portland cement [1]. This fact is due to the development

---

J.M. Ortega • J. Sánchez (✉) • M.A. Climent
Departament d'Enginyeria de la Construcció, Obres Públiques i Infraestructura Urbana,
Universitat d'Alacant (SPAIN)
e-mail: jm.ortega@ua.es; javiersm@ietcc.csic.es

of the hydration reactions of the slag. These reactions densificate the pore structure of concrete due to the formation of additional CSH phases, so the pore size distribution is shifted toward finer pores (pore refinement) [2]. Durability properties of cementitious materials are known to be directly related to the microstructure of these materials [3, 4, 5]. Then the use of slag cements also reduces the permeability of concrete and improves their resistance to aggressive ions ingress, even under non-optimal laboratory conditions at early hardening ages [6].

Real concrete structures will usually harden in different environmental conditions depending on their geographical location. The different temperature [7] and especially the different relative humidity [8] present in the environment may influence the development of hydration reactions of slag and the service properties of concretes and mortars made with slag cements. This influence may cause a different microstructure [9] and, as a consequence, different service properties of these concretes and mortars, such as the diffusion coefficient of chlorides [10] and compressive strength [11]. In this work, mortars made with two different cement types, an ordinary Portland cement and a ground granulated blast-furnace slag cement, were tested in laboratory conditions and at a low relative humidity. The development of their microstructure and their changes in service properties were studied at different ages until 90 days, as a function of the relative humidity of the environment.

2 Experimental Setup

2.1 *Sample preparation*

Mortar samples were prepared using an ordinary Portland cement (OPC), CEM I 42.5 R (referred to as CEM I from now on), and a ground granulated blast-furnace slag (GGBS) cement (with a content of GGBS between 66-80% of total binder), III/B 42.5 L/SR (referred to as CEM III from now on), according to the Spanish standard UNE EN 197-1. Two different water to cement ratios, 0.4 and 0.5, were employed. Fine aggregate was used according to the Spanish standard UNE EN 196-1. The aggregate to cement ratio was 3:1 for all the mortars. Two kinds of samples were prepared. One set was cast in cylindrical moulds of 10 cm diameter and 15 cm height, while the rest of the samples was cast in prismatic moulds of dimensions 4 cm x 4 cm x 16 cm according to the standard UNE EN 197-1. Samples were kept in a 95% RH chamber at 20°C for 24 hours. Thereupon they were demoulded and cylindrical samples were cut into slices of 5 cm thickness.

Constant environmental conditions (RH and temperature) were maintained with glycerol solutions introduced in hermetically sealed containers. Solutions were prepared according to the German standard DIN 50 008 part 1. Containers were introduced into a chamber with controlled temperature. Two different but both constant environmental conditions were distinguished: condition A (100% RH and 20°C, optimum laboratory condition) and condition B (65% RH and 20°C). Condition B is very similar to environmental conditions of Alicante, in Southeast of Spain, with a Mediterranean climate. Tests were performed at 7, 28 and 90 days of age.

2.2 Mercury intrusion porosimetry

In order to study the microstructure of mortar samples, mercury intrusion porosimetry was used. This is a well-known and often used technique [12], in spite of the major problems reported of employing it [13, 14, 15]. Before testing the samples were oven dried for 24 hours at 105°C. Two measurements were made on each sample. The porosimeter employed was an Autopore IV 9500 from Micromeritics. The total porosity, the pore size distribution and also the amount of mercury retained after the end of the experiment were studied. This parameter provides information on the possible tortuosity of pore network [12].

2.3 Capillary absorption test

The capillary absorption test was performed according to the Spanish standard prUNE 83.982. Samples were completely dried for 24 hours in an oven at 105°C, in contrast to Rilem recommendation [16] that suggests a saturation degree of 70% of the samples. The election of complete drying was made in order to accelerate the test and to prevent changes in the microstructure occurring as a result of the hardening environment, especially at early ages. For each cement type, environmental condition and w:c ratio, three different cylindrical samples, with 10 cm of diameter and 5 cm thickness, were tested. The results obtained are the capillary suction coefficient and the water effective porosity according to these expressions:

$$K = \frac{\delta_a \cdot \varepsilon_e}{10 \cdot \sqrt{m}} \text{ with } m = \frac{t_n}{h^2} \quad (1)$$

$$\varepsilon_e = \frac{Q_n - Q_0}{A \cdot h \cdot \delta_a} \quad (2)$$

where ε_e is the effective porosity. Q_n is the weight of the sample at the end of the test, g. Q_0 is the weight of the sample before starting the test, g. A is the surface area of the sample in contact with water, cm². h is the thickness of the sample, cm. δ_a is the density of water, 1 g/cm³. K is the capillary suction coefficient, kg/m²min$^{0.5}$. m is the resistance to water penetration by capillary suction, min/cm². t_n is the time necessary to reach the saturation, minutes.

2.4 Steady-state diffusion coefficient obtained from saturated sample's resistivity

Electrical resistivity measurement of mortars or concretes is an indirect method to study their pore connectivity and to estimate their steady-state diffusion coefficient (D_S). Resistivity was determined from electrical resistance of saturated samples. Electrical resistance of a saturated sample was determined using impedance spectroscopy [4, 8]. Impedance measurements were carried out using the impedance analyzer Agilent 4294A. Impedance spectra were obtained in the frequency range from 100 Hz to 100 MHz, using a contacting method, with direct contact of the electrode with the sample. Impedance spectra were fitted with the equivalent circuit proposed by Cabeza et al. [12]. This circuit includes two times constants. One of the circuital elements is the resistance R_1, and this is associated with pores which cross the sample [12]. This resistance R_1 is equivalent to the electrical resistance of sample. The sample's steady-state diffusion coefficient (D_S) was calculated according to the expression [17]:

$$D_S = \frac{2 \cdot 10^{-10}}{\rho} \quad (3)$$

where ρ is the electrical resistivity of the sample, $\Omega \cdot m$, calculated from electrical resistance of saturated sample.

Samples were saturated for 24 hours according to ASTM Standard C1202-97 [18]. These saturated samples were later used for the forced migration tests. For each cement type, environmental condition and w:c ratio, three different samples were tested.

2.5 Forced migration test

The forced migration test was performed according to Scandinavian standard NT Build 492 [19]. The result of the test is the non-steady-state chloride migration coefficient (D_{NTB}), and it can be calculated according to this expression:

$$D_{NTB} = \frac{0{,}0239 \cdot (273+T) \cdot L}{(U-2) \cdot t} \left(x_d - 0{,}0238 \sqrt{\frac{(273+T) \cdot L \cdot x_d}{U-2}} \right) \quad (4)$$

where: D_{NTB} is the non-steady-state migration coefficient, x 10^{-12} m^2/s, U is the absolute value of the applied voltage, V. T is the average value of the initial and final temperatures in the anodic solution, °C. L is the thickness of the sample, mm, and t is the test duration, hours.

Samples were saturated for 24 hours before the test according to ASTM Standard C1202-97 [18]. For each cement type, environmental condition and w:c ratio three different cylindrical samples, with 10 cm of diameter and 5 cm thickness, were tested.

2.6 Mechanical strength test

The compressive and flexural strengths were measured according to the Spanish standard UNE EN 196-1. Three different prismatic samples were tested for each cement type, environmental condition and w:c ratio.

3 Results and discussion

3.1 Mercury intrusion porosimetry results

Total porosity, Hg retained and the pore size distribution of the samples were studied. The results of total porosity are depicted in Figure 1. A. Samples with w:c ratio 0.5 presented higher total porosities than those obtained for samples with w:c ratio 0.4.

For environmental condition A, the total porosity essentially remained constant between 7 and 90 days for samples prepared with CEM I. Only samples with w:c ratio 0.4 have an important porosity decrease between 28 and 90 hardening days. Total porosity of samples of CEM III showed a significant decrease between 7 and 28 days. This decrease only continued between 28 and 90 days for w:c ratio 0.5 samples. These results could mean that a higher RH accelerates the development of

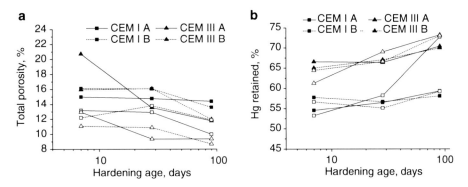

Fig. 1 (a) Evolution with time of total porosity for both cement types and environmental conditions. (b) Results of mercury retained for samples studied. Full symbols are for w:c ratio=0.5, while open symbols are for w:c ratio=0.4

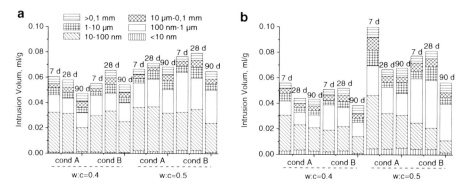

Fig. 2 (a) Pore size distribution for CEM I samples under conditions A and B. (b) Pore size distribution for CEM III samples under conditions A and B

the hydration reactions, especially the hydration of slag. With these reactions, new solids appear and total porosity decreases at early ages.

For environmental condition B, total porosity of CEM I samples and w:c ratio 0.5 kept constant until 28 days and decreased between 28 and 90 days. Total porosity in CEM III samples remained constant between 7 and 28 days, and decreased between 28 and 90 days. A possible explanation is that the lower RH slows down hydration reactions, and the decrease of total porosity happens later.

The study of the mercury retained in the sample after the end of the experiment enables to obtain information on the possible tortuosity of the pore network. The results of the amount of mercury retained are depicted in Figure 1.B. The percentage of mercury retained increased for most of the samples studied. For samples cured under condition A, this increase of mercury retained is higher than that corresponding to samples under condition B. These results could mean that a higher RH helps the hydration reactions of clinker and slag. Then new solids are quickly made and tortuosity of pore network increases.

The study of pore size distribution of samples was done considering the following diameter ranges: < 10nm, 10-100 nm, 100 nm-1 μm, 1-10 μm, 10 μm-0.1 mm and > 0.1 mm. For all the samples studied, the majority ranges were 10-100 nm and 100 nm-1 μm. Values of the contributions to total porosity are shown in Figure 2.A for samples prepared with CEM I. The intrusion volume decreased with age for these mortars. At 7 days of age, the intrusion volume is higher for CEM I samples under condition B. However, this volume is similar for the majority of the samples at 90 days of age, independently of environmental conditions. The pore volume of a majority of ranges decreased with time. From these results, it can be concluded that the higher RH accelerates the hydration reactions of CEM I samples, and then the pore volume of the majority of ranges decreases mainly at early ages. For samples prepared with CEM III, the pore size distribution is shown in Figure 2.B. The intrusion volume decreased quickly for samples under condition A. For environmental condition B, the intrusion volume decreased at 90 days of age. These results show that a high RH in the environment makes the development of hydration reactions

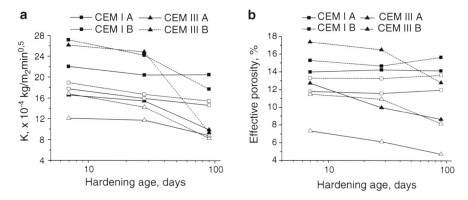

Fig. 3 (**a**) Variation of the capillary suction coefficient with time for both cement types and conditions. (**b**) Results of effective porosity. The symbols have the same meaning as in Figure 1

of slag easier, and then the intrusion volume decreases quickly. Nevertheless, if RH is lower, the intrusion volume decreases later in the case of the cement type III that contains ground granulated blast-furnace slag.

3.2 Capillary absorption results

The capillary suction coefficient (K) and effective porosity of samples are obtained from the tests. The results of capillary suction coefficient are shown in Figure 3.A. For environmental condition A, coefficient K decreased with time for CEM I and III samples. Under environmental condition B, coefficient K also decreased with age for CEM I and CEM III samples, but these decreases are observed mainly between 28 and 90 days of age.

The effective porosity of samples hardened in condition A was lower than the porosity of samples cured in condition B. For both environmental conditions, samples prepared with w:c ratio 0.5 had higher effective porosities than those obtained for samples with w:c ratio 0.4. For environmental condition A, samples with CEM III showed a lower effective porosity than those produced with CEM I for the same w:c ratio. As can be seen in Figure 3.B for each cement type, and w:c ratio the values of the water porosity are always greater under condition B. This parameter has the physical meaning of the volume fraction accessible by water, and as a consequence by the aggressive species (i.e. chloride). This result proves that the laboratory conditions are optimal, and give a greater durability, from this point of view. This behaviour can be explained in terms of the higher RH present in the environment A. The unrestricted presence of water in the environment makes the development of the hydration reactions easier, especially of slag. The products of these reactions are solid phases that cause a more compact structure of the materials.

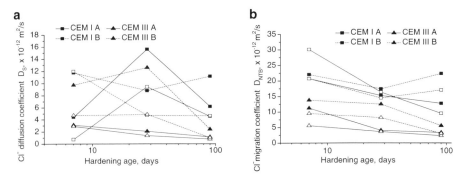

Fig. 4 (**a**) Results of the steady-state diffusion coefficient (D_S) for both types of cements studied. (**b**) Results of the non-steady-state migration coefficient (D_{NTB}). The symbols have the same meaning as in Figure 1

When the RH is less than 100% (65% RH for condition B) these reactions are slower, especially the hydration reactions of slag. This fact is reflected in the results of the effective porosity. As can be seen, the influence of the relative humidity has a greater importance for CEM III. Samples hardened at 100% RH show a decreasing tendency from the beginning of the study, while samples hardened at 65% RH show an important decrease of the effective porosity between 28 and 90 days. From these results it can be concluded that the lower relative humidity causes a slower development of the hydration reactions of slag, meaning that until 90 days a worse behaviour can be expected from the point of view of the aggressive ingress in concrete. The high decrease of the effective porosity from 28 to 90 days for the CEM III suggests a study at later ages, when the behaviour of cement with slag can be similar in both environments. The behaviour of CEM III is always better than CEM I, even for the hardening condition with less water, the difference in the effective porosity increases with time, which means a big improvement with time of the durability of mortars containing CEM III.

3.3 Steady-state diffusion coefficient from resistivity results

The results of the steady-state chloride diffusion coefficient (D_S) are shown in Figure 4.A. Under environmental condition A, D_S had lower values for CEM III samples than for CEM I. For samples prepared with CEM III, D_S decreased with time. This coefficient increased until 28 days for CEM I samples, and decreased since then until 90 days. For hardening condition B, CEM III mortars had smaller values of chlorides diffusion coefficient at greater ages. For CEM I samples, D_S decreased between 7 and 28 days of age, and remained practically constant until 90 days hardening. This coefficient decreased from 28 days until 90 days of age for samples made with CEM III. As could be expected from previous results, the results of steady-state diffusion coefficient could mean that a higher RH facilitates the hydration reactions, especially in the case of CEM III samples. Then new solids are produced faster and the steady-state diffusion coefficient decreases at early ages. When RH in the environment is lower, the development of hydration reactions is slower, and then the steady-state diffusion coefficient decreases later.

3.4 Forced migration results

The results of the non-steady-state migration coefficient (D_{NTB}) are depicted in Figure 4.B. The migration coefficient has smaller values for samples prepared using CEM III than those corresponding to CEM I. This result could be expected from the results obtained for the total and water effective porosities. For mortar samples with CEM I under environmental condition A (100% RH), the samples with w:c ratio 0.4 had the highest migration coefficient at 7 days age. The chloride migration coefficient obtained with these samples decreases with the hardening age, and at the age of 90 days their migration coefficient is the lowest of all the CEM I samples.

For environmental condition B (65% RH), samples of OPC and w:c ratio 0.5 had higher D_{NTB} than samples with w:c ratio 0.4, and both presented a similar decreasing tendency of D_{NTB} between 7 and 28 days age. As should be expected from previous results, samples prepared with CEM III in condition A had an important decrease of D_{NTB} with special importance between 7 and 28 days. In environmental condition B, the biggest decrease of the migration coefficient occurred between 28 and 90 days. At the age of 90 days, D_{NTB} was very similar for all the samples of CEM III. Again, it is shown that a high RH in the environment makes easier the development of hydration reactions, mainly of slag, and then, the non-steady-state migration coefficient decreases quickly. Nevertheless, if the RH is lower, these reactions develop slower and the non-steady-state migration coefficient decreases later, but also reaches good values for CEM III.

3.5 Mechanical strength results

Compressive and flexural strengths were studied. Compressive strength results are depicted in Figure 5.A. Flexural strength results are shown in Figure 5.B. Both strengths increased for the majority of the studied samples. For samples hardened under condition A, this increase is higher than that obtained under condition B. These results could mean that a higher RH accelerates the hydration reactions of clinker and slag. Then new solids are quickly formed and the gain of strength is faster in this case, in agreement with all the results already discussed.

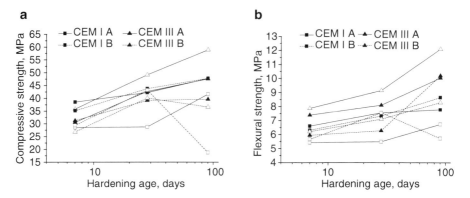

Fig. 5 (**a**) Evolution of the compressive strength with the age of mortar samples. (**b**) Results of the flexural strength for samples studied. The symbols have the same meaning as in Figure 1

4 Conclusions

The main conclusions that can be obtained from the results previously discussed can be summarized as follows:

- Relative humidity has an influence on materials properties. This influence is qualitatively similar for both types of cements studied.
- In general, slag cement samples show better properties of durability at 90 days of age.
- Compressive strength increased with age for the majority of samples studied.
- The improvement of the durability properties is delayed in environments with relative humidity less than 100%. However, after 90 days hardening these properties have reasonable good values.
- Slag cement hardened under environmental conditions of Mediterranean climate, has good service properties at early ages hardening, even better than Portland cement.

Acknowledgements This work has been financially supported by the Ministerio de Ciencia e Innovación of Spain though projects BIA2009-07922 and BIA2010-20548. J.M. Ortega is indebted to the Ministerio de Educación of Spain for a fellowship of the "Formación Personal Investigador (FPI)" programme (reference BES-2008-002650). Authors would like to thank Cementos Portland Valderribas, S.A. and Holcim España, S.A. for providing the cements studied in this work.

References

[1] Bijen, J. (1996). *Constr. Build. Mater.*, vol. 10, p. 309–314.
[2] Manmohan, D., and Mehta, P.K. (1981). *Cement, Concrete, and Aggregates*, vol. 3, n. 1, p. 63–67.

[3] Sánchez, I., López, M.P. and Climent, M.A. (2007). In: *Durability and Degradation of Cement Systems: Corrosion and Chloride Transport*, Proceedings of the 12th International Congress on the Chemistry of Cement, vol. T4.04-4, Beaudoin, J.J., Makar, J.M. and Raki, L. (Eds.), National Research Council of Canada, Montreal.
[4] Sánchez, I., López, M.P., Ortega, J.M. and Climent, M.A. (2011). *Mater. Corros.*, vol. 62, p. 139–145.
[5] Sánchez, I., Albertos, T.S., Ortega, J.M. and Climent, M.A. (2010). In: *Proceedings of the 2nd International Conference on Sustainable Construction Materials and Technologies*, vol. II, p. 655-666, Zachar, J., Claisse, P., Naik, T.R. and Ganjian, E. (Eds.), UWM Center for By-Products Utilization, Ancona (Italy).
[6] Ortega, J.M., Ferrándiz, V., Antón, C., Climent, M.A., and Sánchez, I. (2009). In: *Materials Characterisation IV: Computational Methods and Experiments*, p. 381–392, Mammoli, A.A. and Brebbia, C.A. (Eds.), WIT Press, New Forest (UK).
[7] Schindler, A.K. (2004). *ACI Mat. J.*, vol. 101, n. 1, p. 72–81.
[8] Ortega, J.M., Sánchez, I. and Climent, M.A. (2010). In: *Proceedings of the 2nd International Conference on Sustainable Construction Materials and Technologies*, vol. I, p. 277–287, Zachar, J., Claisse, P., Naik, T.R. and Ganjian, E. (Eds.), UWM Center for By-Products Utilization, Ancona (Italy).
[9] Escalante-García, J.I. and Sharp, J.H. (2001). *Cem. Concr. Res.*, vol. 31, p. 695–702.
[10] Ramezanianpour, A.A., and Malhotra, V.M. (1995). *Cem. Concr. Comp.*, vol. 17, p. 125–33.
[11] Ezziane, K., Bougara, A., Kadri, A., Khelafi, H. and Kadri, E. (2007). *Cem. Concr. Comp.*, vol. 29, p. 587–593.
[12] Cabeza, M., Merino, P., Miranda, A., Nóvoa, X.R. and Sánchez, I. (2002). *Cem. Concr. Res.*, vol. 32, p. 881–891.
[13] Diamond, S. (1999). *Cem. Concr. Res.*, vol 29, p. 1181–188.
[14] Diamond, S. (2000). *Cem. Concr. Res.*, vol 30, 1517–1525.
[15] Gallé, C. (2001). *Cem. Concr. Res.*, vol. 31, p. 1467–1477.
[16] Rilem recommendation TC 116-PCD (1999). *Mater. Struct.*, vol. 32, p. 174–179.
[17] Andrade, C., Alonso, C., Arteaga, A., and Tanner, P. (2000). In: *Proceedings of the 5th CANMET/ACI International Conference on Durability of Concrete*, Supplementary papers, p. 899-915, Malhotra, V.M. (Ed.), American Concrete Institute, Barcelona (Spain).
[18] ASTM (1997). "*ASTM Standard C1202-97: Standard test method for electrical indication of concrete's ability to resist chloride ion penetration.*"
[19] Nordtest (1999). "*NT Build 492. Concrete, mortar and cement-based repair materials: Chloride migration coefficient from non-steady-state migration experiments.*" Espoo (Finland).

Study of corrosion rate variability in indoor and outdoor specimens

Filipe Pedrosa and Carmen Andrade

Abstract Behavior, time evolution and variability of the corrosion process in reinforced concrete structures is an important factor to have in mind if structural service life is to be determined. Corrosion development and its consequences over time are often empirically deduced. It would be significant to improve as much as possible the prediction accuracy of corrosion process behavior.

In the present paper, a series of hygrothermal and corrosion data obtained for indoor and outdoor reinforced concrete specimens are summarized by means of descriptive statistics.

For a T-beam exposed to continental weather conditions, temperature, relative humidity, corrosion potential and corrosion current density measurements were taken since the year of 1993. The same measurements were made for an outdoor pile exposed to the same conditions. The indoor studied specimens are two small slabs exposed to lab conditions for which corrosion measurements were taken in the first 96 days period. Chloride was added to the concrete mixture of one specimen in order to accelerate the corrosion process.

Electrochemical corrosion rate measurements using modulated confinement of the current were taken by means of Gecor 8 system device. Temperature and relative humidity were measured both inside and outside of the outdoor specimen using a probe.

Hygrothermal and corrosion results were plotted on time scale in order to visualize possible correlations.

A first approach to a statistical analysis is made by describing the raw data obtained by means of average value and standard deviation for different variables. Results show that the accumulated corrosion progress linearly in time. In the outdoor exposed specimen two corrosion rates *Icorr* were noticed, being the highest for the first period where no cracks were yet produced. After cracking occurs corrosion rate decreases. No direct relation between RH and *Icorr* was observed.

F. Pedrosa (✉) • C. Andrade
Center of Safety and Durability of Structures and Materials-CISDEM
(CSIC-UPM), Madrid, Spain
e-mail: filipe.pedrosa@ietcc.csic.es

1 Introduction

There are several theories and models in order to predict the service life of corroding structures [1]. One of the most generally accepted is the conceptual model presented by Tuutti [2], as illustrated in Figure 1, that basically describes two periods, a first one of initiation in which external aggressive agents enter the concrete triggering the deterioration process, and a second one, of propagation, in which this deterioration continues until reaching an unacceptable degree of corrosion.

For the propagation stage significantly less literature exists than for the initiation period and its behavior is often described based on assumptions.

Data on corrosion rate values measured on-site in real size concrete structures are scarce, while the data bank of values obtained for laboratory specimens is relatively large [4]. This constitutes one of the factors causing the present imprecision characterizing the propagation period.

Corrosion and hygrothermal records should be promoted in real size concrete structures on outdoors as well as for indoor specimens, in order to develop a analysis methodology of the data and study their dependence on climatic events but also on the approach to record and represent data in engineering terms. In the present paper results are given for concrete elements exposed in outdoor and indoor conditions and the analysis is made in order to relate the corrosion rates measured to the RH or to temperature.

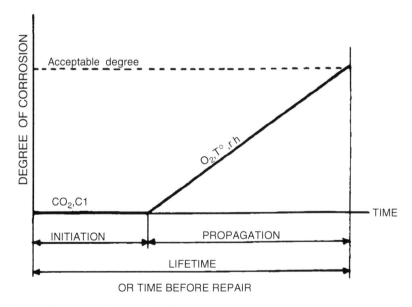

Fig. 1 Tuutti´s model, as illustrated in [3]

2 Experimental

2.1 Specimens

The samples used were: a reinforced concrete T-beam [4] and a square section pile, as illustrated in Figure 2. Both specimens have been contaminated with 3% chloride by weight of cement added to the mix in the form of $CaCl_2$ to promote corrosion and were exposed to Madrid climate. The specimens are 2 meters in length and are situated in the institute´s garden.

The two indoor specimens are two small slabs (50×40×10 cm) of reinforced concrete, one without admixtures and the other (Figure 3) with 3% chloride content added to the mix in the same way as the outdoor specimens. These two slabs were held in laboratory conditions but maintained wet by covering them with a sponge in order to maintain moisture conditions favorable to the corrosion process.

2.2 Techniques

In order to obtain the corrosion potential (*Ecorr*), the resistivity (ρ) and the corrosion rate (*Icorr*), a portable corrosion rate meter, Gecor 8, was used, having modulated confinement of the current. The reference electrode used is $Cu/CuSO_4$ [5, 6].

The relative humidity, and temperature, T, were recorded in the outdoor samples in cavities of the concrete especially made for this purpose. For these measurements, a Vaisala portable hygrometer (Figure 4) has been used [4, 7].

In the beam the data bank obtained is analyzed from 1993 and for the pile from February 2009. Measurements made on the indoor slabs have been recorded for a period of 96 days.

Fig. 2 Reinforced concrete T-beam and pile samples

Fig. 3 Slab with added chloride used in the experimentation

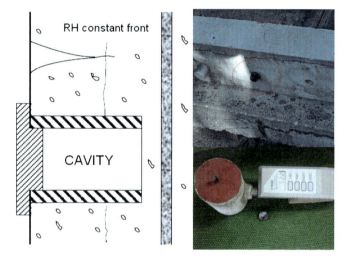

Fig. 4 Vaisala hygrometer used for measurement of the interior RH

3 Results

The temperature inside the cavities of the outdoor specimens (*T*-IN) and temperature in the external atmosphere (*T*-EXT) behave similarly (Figure 5). This is interpreted as due to the metallic tube inserted in the cavity that communicates thermally the exterior environment with the bottom of the cavity.

As for the relative humidity, this is not the case, as RH values remain more stable inside the specimens than in the outdoor environment, as reported in [4, 7]. However, for the last period of recorded time, when the concrete is already severely cracked due to corrosion, this does not occur, as the two environments seem to be now in more direct contact. This is illustrated in Figure 6.

In Figure 7 are shown the corrosion rate values for the outdoor specimens. It is noticeable for the T-beam that *Icorr* fluctuates significantly due to variation of temperature and rain periods. Another remark to make is the much higher values of *Icorr* registered in the first period lasting around one and a half years. After that,

Variability of corrosion rate in indoor and outdoor specimens

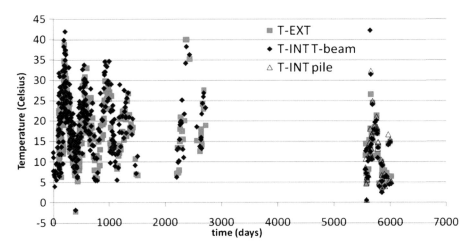

Fig. 5 Temperature evolution in the environment and inside the specimens

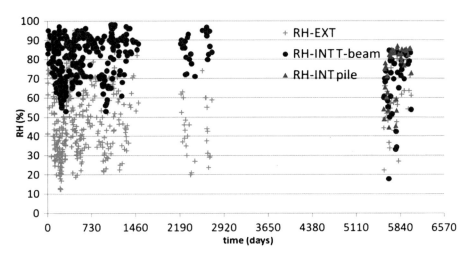

Fig. 6 Evolution of RH in the atmosphere ambient and inside the T-beam

Icorr values decrease greatly and although fluctuating, they remain below 0.2-0.3 μA/cm² in the frontier of negligible-active corrosion ranges.

As for the results in the pile recorded from 2009, the *Icorr* values are in the same range than the initial ones in the beam. No later decrease in the values is noticed.

In Figure 8 are given the *Icorr* values recorded in the indoor slabs. The slab without admixtures presents at the beginning values of *Icorr* greater than 0.1 μA/cm² during the formation of the passive film. Then the steel is passivated and the *Icorr* remains below 0.1 μA/cm². The slab containing $CaCl_2$ gives much higher *Icorr* values, also greater at the beginning and then decreasing to corrosion rates higher than 0.2 μA/cm². This decrease is attributed to the progressive hardening and increase of resistivity.

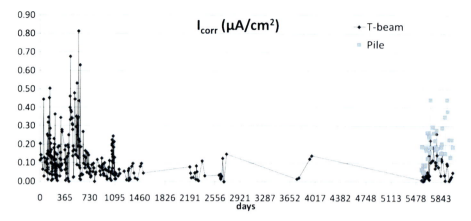

Fig. 7 Evolution of corrosion current density for the outdoor specimens

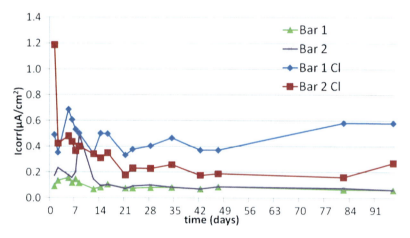

Fig. 8 Evolution of the corrosion current density for the slabs

4 Discussion

Without admixtures, in spite of the wet conditions, the steel passivates and soon shows *Icorr* values smaller than 0.1 µA/cm² defined as negligible corrosion. In presence of $CaCl_2$ several aspects should be commented: 1) the similar trend of *Icorr* values, being around 0.5 ± 0.2 µA/cm² in the initial period, to later decrease to values that are lower in the case of outdoor conditions; 2) these lower values in the outdoor specimens are attributed not only to the increase in resistivity typical of pore refinement but also to the cracks produced due to the corrosion process. The cracking parallel to the reinforcement seems to allow more rapid drying than the uncracked cover. This is confirmed with the case of the pile which exhibits higher corrosion rate values due to the smaller cracking because of higher concrete cover depth.

Variability of corrosion rate in indoor and outdoor specimens

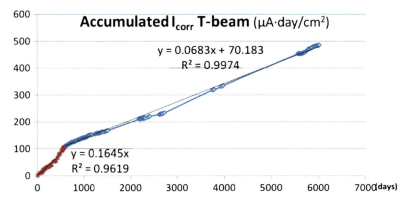

Fig. 9 Accumulated corrosion for the T-beam

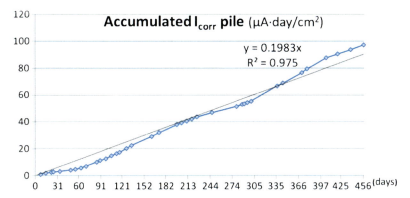

Fig. 10 Accumulated corrosion for the pile

We can see in Figure 9 two distinct periods in the accumulated corrosion evolution. The first one, until the second year, where is noticeable a bigger slope, reverting later to lower values in a second one. A hypothesis that may be formulated to the existence of these two periods is that after a certain level of cracking, corrosion levels drop down, probably due to the consequent effect of drying.

Plotting the accumulated corrosion for the data series of the outdoor pile (Figure 10), being the time period under consideration smaller; we can depict the seasonal effect of the hygrothermal factors that affect corrosion, as the plot has a wave form.

According to existing records the cracking width level of the pile is much lower actually than the one of the beam, being comparable to the state of the T-beam before the first two years of measurement records. This goes in line with the previously formulated hypothesis because the slope of the accumulated corrosion of the pile is similar with the slope observe for the first two years of the T-beam.

As said before in [4] it does not appear to exist any direct relation between the RH values and the corrosion rates obtained. However, by observing Figure 11, it is

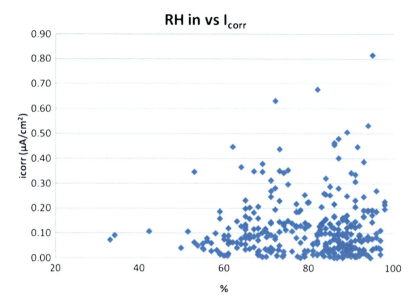

Fig. 11 Internal relative humidity versus current density (T-beam)

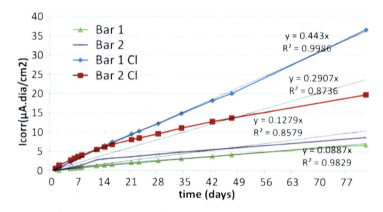

Fig. 12 Accumulated corrosion for the slabs

clearly visible that the inside RH influences the possible maximum values for I$corr$. As RH grows it allows corrosion current density to reach higher values. Relative humidity influences the range of I$corr$ values possible.

Plots were also obtained relating the temperature with I$corr$ and the E$corr$. Very disperse clouds of data points were obtained and no direct relation was visible. The relative humidity is a much more relevant hygrothermal factor influencing corrosion evolution in outdoor environments than the temperature.

The accumulated corrosion for the first 96 days of the slabs is showed in Figure 12.

We see that the slopes for the two bars of the slab containing chlorides are considerably bigger than the ones observed for the outdoor specimen considering full

Table I Summary of descriptive statistics for T-beam (full period of time)

	T-EXT (°C)	T-INT (°C)	RH-EXT (%)	RH-IN (%)	Icorr (µA/cm^2)	log Icorr	Ecorr (mV)
Average value	17.607	18.791	48.328	79.992	0.119	−0.128	−334.0
Std. deviation	7.300	8.043	17.569	12.271	0.119	0.466	107.3
Coef. Variation	41.5%	42.8%	36.35%	15.34%	99.57%	41.3%	32.1%

Table II Summary of descriptive statistics for the pile (full period of time)

	T-EXT (°C)	T-INT (°C)	RH-EXT (%)	RH-IN (%)	Icorr (µA/cm^2)	log Icorr	Ecorr (mV)
Average value	12.01	11.72	58.51	74.83	0.199	−0.759	−346.9
Std. deviation	5.711	6.677	15.185	12.934	0.097	0.238	68.9
Coef. Variation	47.5%	57.0%	26.0%	17.3%	49.0%	−31.3%	19.9%

Table III Summary of descriptive statistics for the slab with added Cl⁻

	Bar 1			Bar 2		
	I$_{corr}$ (µA/cm^2)	log I$_{corr}$	E$_{corr}$ (mV)	I$_{corr}$ (µA/cm^2)	log I$_{corr}$	E$_{corr}$ (mV)
Average value	0.471	−0.337	−423.4	0.352	−0.513	−403.0
Std. deviation	0.106	0.097	35.7	0.238	0.214	133.2
Coef. Variation	22.45%	−28.9%	8.4%	67.58%	−41.79%	28.1%

Table IV Summary of descriptive statistics for the slab without chloride

	Bar 1			Bar 2		
	I$_{corr}$ (µA/cm^2)	log I$_{corr}$	E$_{corr}$ (mV)	I$_{corr}$ (µA/cm^2)	log I$_{corr}$	E$_{corr}$ (mV)
Average value	0.096	−1.04	−187.25	0.143	−0.92	−191.94
Std. deviation	0.029	0.126	62.567	0.104	0.235	56.312
Coef. Variation	30.77%	−12.12%	33.41%	72.80%	−25.67%	29.34%

time scale. It should be in mind that the slab was always kept humid in order to develop corrosion.

Table I and Table II summarize the statistical information for the different parameters measured in the T-be am and the pile exposed to outdoor conditions. I*corr* values present a high coefficient of variation that does not seem to have a direct dependence of the temperature and relative humidity variation. Also it does not seem possible to deduce E*corr* variation from the hygrothermal parameters.

The statistical information obtained for the indoor slabs with and without chloride addition is presented in Table III and Table IV respectively.

A lower variability of I*corr* can be observed when compared to the outdoor beam but a higher one when compared to the outdoor pile. In general, I*corr* values seem to fluctuate considerably.

5 Conclusions

Although much of the data has not yet been thoroughly analysed and more research is needed, some conclusions can be drawn.

It is noticeable that relative humidity on the interior of not seriously cracked concrete exposed to outdoor environment does not vary as much as the relative humidity of the exterior atmosphere.

The hygrothermal parameters influence the range limit of both E*corr* and I*corr*, however not their variation. RH-INT is the most relevant limiting factor.

I*corr* values fluctuate considerably, either in outdoor an indoor specimens. In order to study corrosion behavior over large periods of time the accumulated corrosion rate is a much better alternative way of visualizing the deterioration process.

Corrosion seems to develop differently after a certain level of cracking has been reached. It attenuates with time due to the increase of the concrete resistivity and possible more dry conditions when cracking favours it.

References

[1] Andrade, C., M.C. Alonso, and J.A. Gonzalez, *An Initial Effort to Use the Corrosion Rate Measurements for Estimating Rebar Durability*, in *Corrosion rates of steel in concrete*, ASTM, Editor. 1990: Philadelphia. p. 29–37.
[2] Tuutti, K., *Corrosion Of Steel In Concrete*. 1982, Swedish Cement and Concrete Institute. p. 17–21.
[3] López, W., J.A. González, and C. Andrade, *Influence of temperature on the service life of rebars*. Cement and Concrete Research, 1993. **23**(5): p. 1130–1140.
[4] Andrade, C., C. Alonso, and J. SarrIa, Corrosion rate evolution in concrete structures exposed to the atmosphere. Cement and Concrete Composites, 2002. **24**(1): p. 55–64.
[5] Andrade, C. and C. Alonso, *On-site measurements of corrosion rate of reinforcements*. Construction and Building Materials. **15**(2-3): p. 141–145.
[6] Andrade, C. and C. Alonso, *Test methods for on-site corrosion rate measurement of steel reinforcement in concrete by means of the polarization resistance method*. Materials and Structures, 2004. **37**(9): p. 623–643.
[7] Castillo, Á., *Modelado del efecto de las acciones climáticas en la cantidad de agua en estado líquido del hormigón para la estimación de la corrosión de la armadura*, in *Civil Engineering: Hydraulics and Energy*. 2005, E.T.S.I. Roads, Canals and Ports (UPM). p. 266.

From corrosion rate to accumulated corrosion depth or loss in cross section of reinforcements

Nuria Rebolledo and Carmen Andrade

Abstract In marine environments, reinforcement corrosion develops due to the penetration of chlorides through the concrete pores. Initially the corrosion appears as localized attack, but when the chlorides arrive in high quantities, it extends to the bar surfaces which corrode entirely. In numerous previous studies, the detection of corrosion and its evolution have been monitored by means of measuring and presenting the variation of corrosion rate over time. This representation can be used to deduce when corrosion initiates on a steel bar and how it evolves with time. However, as the corrosion rate changes with temperature and the corrosion process itself evolves, this representation may not be clear enough to be used for comparative purposes. The present work proposes a parallel use of accumulated corrosion depth, which is obtained by the integration of each age of the corrosion rate-time curve. This kind of representation enables the determination of corrosion depth at each age and appears more appropriate for comparative purposes. The procedure is applied to concrete specimens that have been in contact with natural sea water from the Mediterranean Sea for 20 years. The specimens are prepared with different binders in different proportions for purposes of comparison.

1 Introduction

Reinforced concrete placed in a marine environment may suffer premature reinforcement corrosion due to chloride ions penetrating through the concrete pores in time periods shorter than the planned service life of the structure. The corrosion

---

N. Rebolledo (✉)
Center for Safety and Durability of Structures and Materials,
CISDEM (IETcc) CSIC-UPM, Madrid, Spain
e-mail: nuriare@ietcc.csic.es

C. Andrade
Affiliation or Safety and Durability of Structures and Materials,
CISDEM (IETcc) CSIC-UPM, Madrid, Spain

mechanism causes initial pitting corrosion of steel (1–3) inducing a local disruption of the passive layer of steel (4). After this initial depassivation, new pits are generated and a more general attack is produced due to higher chloride amounts reaching the steel surface (5).

While a lot of literature is available on the initiation period for steel corrosion, the propagation of corrosion is much less studied with very few papers reporting on the long term monitoring of corrosion with time.

This paper presents data regarding the corrosion rates, monitored for 20 years, of differentes types of steel bars embedded in concrete specimens fabricated with different types of cement and concrete dosages concrete. This work enables us to comment on the evolution of corrosion during the depassivation period. It has been used to illustrate the need to represent not only the corrosion rate, but also the accumulated corrosion depth as a better means of studying corrosion behavior.

2 Experimental

2.1 Materials

Concrete or mortar specimens of size 15 × 20 × 10 cm were fabricated. Each specimen contained ten reinforcing steel bars, 8 mm in diameter, embedded at various depths in order to have different cover depths. Bare and galvanized steels were embedded at a cover depth of 0.5, 1.5, 2.5, 4 and 7.5 cm. Figure 1 shows the geometry of the specimen used and the depth of each reinforcing steel bar.

To enable the penetration of chlorides only through the front and back surfaces, the lateral surfaces of the specimens were covered first with an epoxy resin and then, with wax (Figure 1).

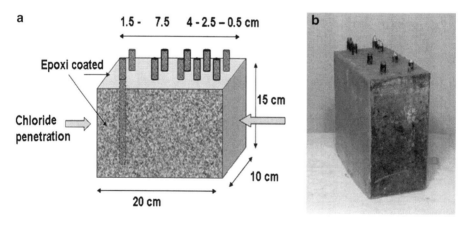

Fig. 1 a) Dimensions of the concrete specimens and cover of the reinforcements; b) Picture of the specimens. Lateral surfaces are covered with epoxy and wax

Table 1 Dosage of the concrete

Type of cement	Cement Content (kg/m³)	w/c	% NO₂⁻ weight cement	Type of steel
Ordinary Portland cement	300	0,6	- - -	Carbon Steel
		0,62	2	
		0,6	- - -	Galvanized Steel
	400	0,6	- - -	Carbon Steel
		0,52	2	
		0,6	- - -	Galvanized Steel

The top and bottom of the embedded reinforcement bars were protected with an electrical insulating tape such that the steel surface in contact with concrete was 23 cm².

The concrete specimens (Table 1) were cured in a chamber at 100% =relative humidity for 60 days. Then they were immersed in tap water for 30 days to saturate the pores. Finally, the specimens were introduced in a tank with natural sea water from the Mediterranean Sea. The specimens were immersed to a level 2 cm from the upper border of the tank to keep the concrete continuously wet. The water was renewed every 60 days. The mean content of chloride in the water was 38 g/l. The temperature ranged from 14°C during winter up to 24°C during summer. The mean oxygen content of the sea water was 5 ppm.

2.2 *Electrochemical techniques and calibration to gravimetry*

The corrosion potential E_{corr} and corrosion rate I_{corr} were monitored during the experiment. The E_{corr} was recorded using a calomel electrode. The I_{corr} was determined through measurement of the polarization resistance R_p (6). A value of 26 mV was used for the B constant (6–7). The corrosion onset was considered when the corrosion rate was higher than 0.1 µA/cm². Measurements were performed at shorter intervals in the beginning, being more differentiated at later time.

Some of the specimens were broken and the bars removed, cleaned and weighed in order to compare the electrochemical losses and the gravimetric ones (7).

3 Results

The evolution of the corrosion potential, E_{corr}, of specimens vs. time is demonstrated in Figure 2. Each figure represents the bars embedded at the 5 different concrete covers tested.

Figure 3 shows the corrosion rate for different specimens. From the representations it is possible to deduce that after curing, the reinforcing steel bars are in a passive state showing $I_{corr} < 0.1 \mu A/cm^2$. The depassivation of the reinforcements is identified when $I_{corr} > 0.1 \mu A/cm^2$. Once the shift in I_{corr} is registered, it evolves through processes of

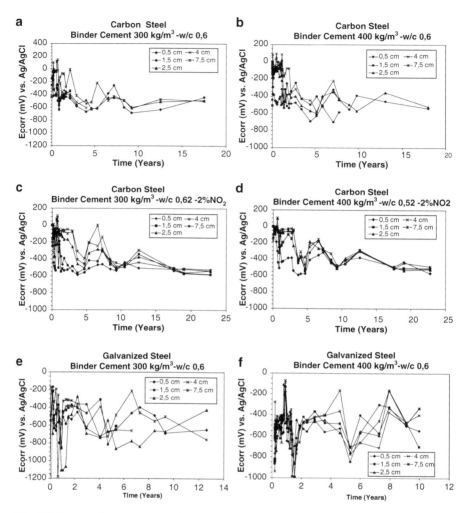

Fig. 2 Influence of concrete cover in potential corrosion (vs. SCE) evolution of reinforcement embedded in **a**) OPC, binder cement 300 kg/m³ and w/c 0.6 (Carbon Steel, CS); **b**) OPC, binder cement 400 kg/m³ and w/c 0.6 (CS); **c**) OPC, binder cement 300kg/m³, w/c 0.62, 2% NO_2 (CS); **d**) OPC, binder cement 400kg/m³ and w/c 0.52 and 2% NO_2 (CS); **e**) OPC, binder cement 300kg/m³ and w/c 0.6 (Galvanized Steel, GS); **f**) OPC, binder cement 400kg/m³, w/c 0.6 (GS)

active/passive states before attaining the value of $I_{corr} > 0.2\mu A/cm^2$. Thus, from a practical point of view, depassivation is not an instant or short period phenomenon and requires a period of months or even years.

After the corrosion process was continuously established, the I_{corr} increased steadily with time although events of 'higher-lower values' were also noticed. These cycles were attributed to changes in temperature, as the sea-water tank was in a

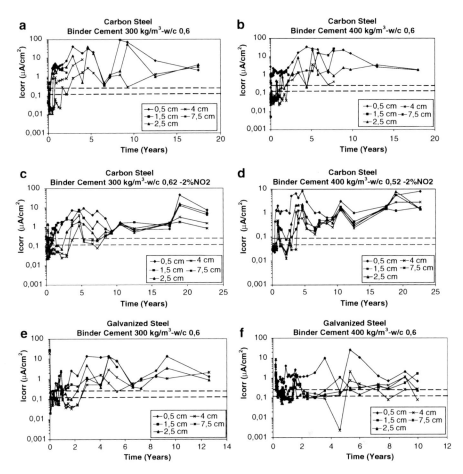

Fig. 3 Influence of concrete cover in corrosion rate evolution of reinforcement embedded in a) OPC, binder cement 300 kg/m³ and w/c 0.6 (CS); b) OPC, binder cement 400 kg/m³ and w/c 0.6 (CS); c) OPC, binder cement 300kg/m³, w/c 0.62 and 2% NO$_2$ (CS); d) OPC, binder cement 400kg/m³ and w/c 0.52 and 2% NO$_2$ (CS); e) OPC, binder cement 300kg/m³ and w/c 0.6 (GS); f) OPC, binder cement 400kg/m³, w/c 0.6 (GS)

non-air-conditioned chamber, and depending upon the season of the year, the temperatures varied from around 14°C to 25°C. These temperatures may justify the changes in values of the corrosion rates measured in the specimens submerged in sea water.

As expected, in all cases, the depassivation times for the reinforcing steel bars due to the action of chlorides, was longer for higher covers. The use of inhibitors increased the depassivation time with respect to OPC. But, after ten years, all covers in all of the specimens were insufficient to prevent the corrosion of reinforcements.

4 Discussion

4.1 Use of the accumulated corrosion depth

The fact that the I_{corr} measured is not constant can be attributed to two factors: certain variations in temperature in the chamber where the tank is placed and the evolution of the corrosion process itself, as the advancement of corrosion produces rust and concrete microcracks which modify the surroundings.

Due to the limitations of interpretation of the corrosion rate parameter in case of localized corrosion, it becomes essential to study the progress of the accumulated corrosion depth, P_x. This is calculated by multiplying I_{corr} with time (preferably in years, for simplicity).

Figure 4 shows the evolution (as if it were homogeneous) of P_x of the bars embedded in the concretes. Now the possibility of comparison is more evident. The effect of the cover thickness as well as the improved performance of lower w/c ratios and higher cement contents is much better appreciated. The significantly favorable behavior caused due to the use of inhibitor (NO_2^-) and galvanized steel is easily identified because of the evident delay in corrosion initiation period. Figure 4 is reminiscent of the typical service life model.

From the slope obtained in each curve of accumulated corrosion, during the propagation period, the average steel section that is lost each year can be obtained (Figure 5).

4.2 Influence of concrete variables in marine environments

The effects of the cover thickness, the use of inhibitors or galvanized steels can be studied from Figure 2 to Figure 5, and are as expeted. The beneficial effects of the addition of 2% NO_2^- in cement weight in the water mix, delay onset of corrosion and reduce the section loss to 10% after 20 years in a sea water environment.

A similar behaviour can be obtained for galvanized steel, where the section loss is around 15% (see Figure 6).

4.3 Reliability of the electrochemical techniques

Figure 7 shows the comparison of the electrochemical and gravimetric losses for the bars embedded in three specimens. The other three specimens still remain submitted to testing. Only very few points were in the limit of the double or half of agreement value and no values were found out of this validity limit. This agreement between gravimetric and electrochemical calculated losses once more reveals the accuracy of the Rp technique and its validity, in particular when referring to the values registered in the cracked cover.

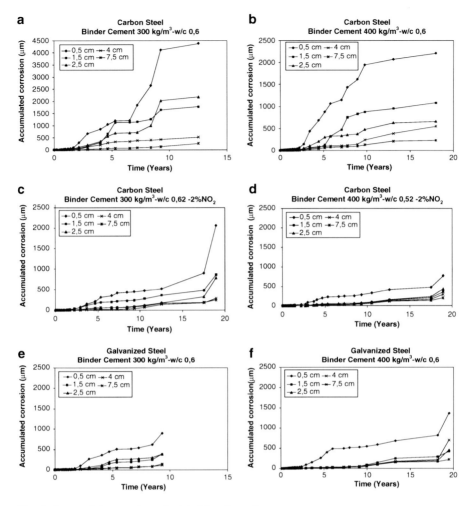

Fig. 4 Accumulate corrosion calculated for all specimens

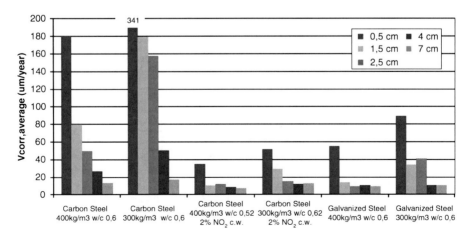

Fig. 5 Average V_{corr} for all specimens

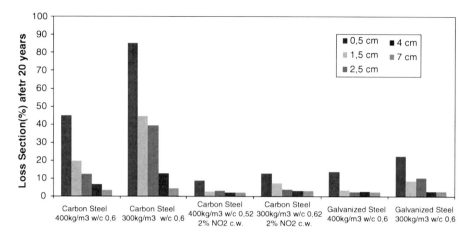

Fig. 6 Percentage of loss section alter 20 years immersion in sea water

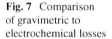

Fig. 7 Comparison of gravimetric to electrochemical losses

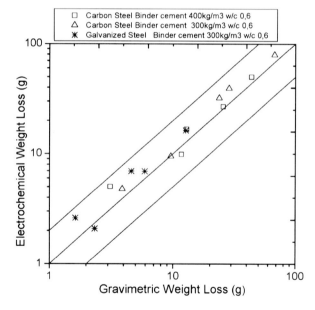

5 Conclusions

- The accumulated corrosion seems a better parameter than the representation of the instantaneous corrosion rate to study the effect of the different concrete variables in long term tests. It also enables the calculation of the residual bar cross section, either homogenously distributed or as localized attack.
- The addition of nitrite has a beneficial effect on the corrosion behavior of steel, as it results in a delay of the depassivation time and a significant reduction

of the corrosion rate. The use of galvanized generates a similar behavior by significantly reducing the losses in steel section.
- As it was to be expected, the binder content and the w/c ratio are also determinant parameters in order to reduce the initial pitting corrosion.

References

[1] Page, C.L., Short N.R. and Holden W.R. (1986), Cem. Concr. Res., vol. 16, n.1, p.79–86.
[2] Gouda, V.K. (1970), Br. Corros., vol 5, p. 198–203.
[3] Tuutti, K. (1982), CBI Publ. Stockholm
[4] Rosenfeld, I.L., Danilov I.S. (1967), Corros. Sci., vol. 7, n. 3, p. 129–142.
[5] Lambert, P., Page C.L., Vassie P.R.W. (1991), Mater. Struct., vol. 24, n. 143, p.351–358.
[6] Andrade C., Alonso C. (1996), Const. Build Mater, vol. 10, n. 5, p.315–328.
[7] Andrade C., Alonso C. et al (2004), Mater. Struct., vol. 37, p. 623–643

Steel Corrosion in a Chloride Contaminated Concrete Pore Solution with Low Oxygen Availability

Lina Toro, Carmen Andrade, José Fullea, Isabel Martínez, and Nuria Rebolledo

Abstract It is commonly mentioned that in concrete chloride induced corrosion is controlled by the oxygen content in such a manner that in water saturated conditions no oxygen will be present and thus no corrosion can develop. In the present paper, experimentation has been made in low oxygen availability "pore" solutions with several amounts of chlorides. These situations may represent the case of a water saturated concrete. The results indicate that at very low oxygen contents, i.e. almost negligible because complete removal is very difficult, corrosion may develop in presence of chlorides. The presence or absence of corrosion is influenced by the amount of chloride, its corrosion potential and the steel surface condition.

1 Introduction

The cathodic reaction in concrete is said to be the reduction of the oxygen that penetrates through the concrete pores in presence of water that takes part in the reaction. Eqn. (1):

$$O_2 + 2\,H_2O + 4\,e^- \rightarrow 4\,OH^- \tag{1}$$

Due to the high alkalinity of the pore solution, the steel bars embedded in concrete will remain passive [1] unless a decrease of the pH of the electrolyte in contact with the metal is produced. This decrease can be induced by the neutralization of the alkalinity by carbon dioxide (carbonation) or by the presence of chlorides which are able to decrease the pH and induce local corrosion and pits [2]. This corrosion is

---

L. Toro • C. Andrade (✉) • J. Fullea • I. Martínez • N. Rebolledo
Center for Safety and Durability of Structures and Materials, CISDEM-Instituto de Ciencias de la Construcción Eduardo Torroja, CSIC-UPM, C/Serrano Galvache n° 4., 28033, Madrid, Spain
e-mail: andrade@ietcc.csic.es

known to be the cause of structural damages and its prevention or reduction is of primary importance.

Several methods can be used to delay the onset corrosion or to try to stop its development [3]. For the corrosion suppression it is necessary to reduce the kinetics of the electrolytic oxidation-reduction process, reducing at least one of the reactions. For the inhibition of the cathodic reaction, the removal of the reagents, water or oxygen, from the electrolyte can be considered as an option. In the literature it is widely accepted that in water saturated concrete the corrosion rate is controlled by oxygen diffusion though the concrete cover [4][5] and thus corrosion cannot develop in the absence of oxygen.

In present paper experiments have been made in chemical solutions simulating that contained within concrete pores with several amounts of chlorides. By bubbling N_2, oxygen has been removed. Interesting results have been obtained when measuring corrosion in these conditions, demonstrating that is possible to have active corrosion due to chloride attack even when the oxygen concentration in the electrolyte is negligible. In these conditions, the presence or absence of corrosion and its typology depends on the amount of chloride in contact with the steel, its corrosion potential and the surface roughness of the steel.

2 Experimental

The experiments were carried up at free corrosion potential simulating the natural corrosion conditions.

2.1 Corrosion cell preparation

As the solutions were very alkaline, polyethylene corrosion cells (250 ml capacity) were used in all the experiments. A simulated concrete pore solution was prepared using de-carbonated water (boiled) saturated in $Ca(OH)_2$ (2.2 g/L). Different amount of chlorides were added depending on each experiment.

The cell schemes for progressive chloride addition and fixed chloride concentrations tests are shown in Figure 1a.

2.2 Test procedures

Two types of tests were performed. In one the chloride concentration was increased progressively over the course of the test, whereas in the other the chloride concentration was kept constant from the beginning.

Steel Corrosion In Deaerated Solutions

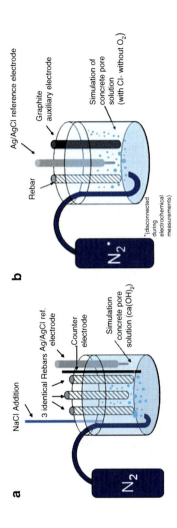

Fig. 1 Cell scheme used for free corrosion potential test. **a)** Progressive chloride addition. **b)** Fixed chloride concentration

Table 1 Summary of the tested performed at free corrosion

Test condition	NaCl content	Test duration	Technique used
Progressive NaCl addition	1 ml NaCl (6 M) every 30 min. until reaching a concentration of 2,5M NaCl in the solution	130 h	E_{corr}, R_p: I_{corr} [O_2] and pH
Fixed NaCl concentration	2,5 M 0,5 M	120 h 700 h	E_{corr}, R_p: I_{corr} Electrochemical weight loss vs gravimetric weight loss.

Progressive chloride addition Three steel bars were introduced in a cell with saturated Ca(OH)$_2$ solution, Figure 1a. The steel area exposed to the electrolyte was 3.77 cm². After 24 hours of N$_2$ bubbling, the O$_2$ concentration in the solution was checked and the value was around 0.4 ppm. At this moment, a progressive addition of NaCl 6M in the cell was started (1 ml each 30 minutes). The test was finished when the chloride concentration in the cell reached the value of 2.5M and the steel exhibited a $I_{corr} > 0,2$ µA/cm², which indicates steel depassivation. The total duration of the experiment was approximately 130 hours, displacing the O$_2$ continuously by the N$_2$ bubbling in the solution.

The electrochemical parameters measured during the test were the corrosion potential and corrosion rate of the three steel bars. The oxygen concentration in the solution was measured at the beginning of the test, using an oxygen meter with an electrode ORION. The pH values were registered over the course of the test by means of an electrode for very high alkalinities and a pH meter ORION 5STAR.

Fixed chloride concentration In this type of test, the chloride concentration was set from the beginning. The cells used are those shown in Figure 1b. Two chloride concentrations were chosen for short term experiments and for the comparison between electrochemical and gravimetric results. Chloride concentrations values of 0.5M and 2.5M NaCl were tested. The experimental procedure was the same in both cases with the difference that the 2.5M NaCl case had a duration of five days and the 0.5M NaCl concentration of one month. E_{corr} and I_{corr} were registered over time.

The conditions for the experiments are summarized in Table 1.

2.3 Electrochemical techniques used

Corrosion Potential measurement, E_{corr}: The potential values were monitored with an AGILENT 34970A datalogger, determining the difference of potential between the steel and the Ag/AgCl reference electrode.

Corrosion Rate measurement, I_{corr}: Polarization Resistance (R_p) method was used. Results of R_p are used to calculate the instantaneous I_{corr}, through the Stern-Geary formula using a constant B = 26mV (I_{corr} =26/R_p) [6]. A potentiostat (AMEL 550)

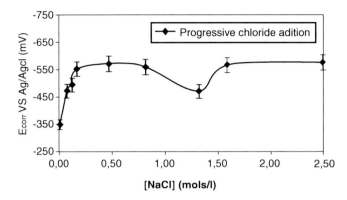

Fig. 2 Corrosion potential (versus Ag/AgCl electrode) monitoring during experiment with progressive chloride addition

with ohmic drop was used. The R_p was measured by linear polarization in a range of ±10 mV centred in the E_{corr} and with a scan rate 10 mV/min.

Potentiostatic test: The potentiostat used for the tests was the AMEL 550. The current flowing between the steel and the counter electrode was monitored during the tests by an AGILENT 34970A datalogger.

3 Results

3.1 Progressive chloride addition

Figure 2 shows an average of the corrosion potential of the three bars (which presented very similar individual values) after each chloride addition. It can be observed that the potential evolves towards more negative values during the progressive increase of chloride addition to the electrolyte. The initial potential of the steels was around -350 mV, being at the end of -570 mV when the NaCl concentration in the electrolyte was 2.5 M. The trend however is not continuous as the potential drops to values around -570mV with small addition of chlorides and after, it shifts to values around to -470mV to finally reach again -570mV.

An average of the corrosion rate variation with the chloride addition for the three steels is presented in Figure 3. The initial values are around 0,08 µA/cm² in solution without chloride. Later, the I_{corr} increase progressively and its value was in the threshold between (0,1 – 0,2 µA/cm²) [8] until the chloride concentration was around 0,8M. When the chloride concentration is higher than 0,9M the I_{corr} shows corrosion rate values over this depassivation threshold. The last I_{corr} value registered was 3.5µA/cm² in solution with 2.5M NaCl.

The evolution of the pH with NaCl addition is shown in Figure 4. There is a sharp initial decrease at about 0.25M chloride concentration. Between 0.20M and 1.5M

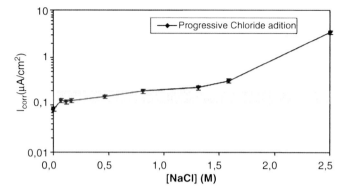

Fig. 3 Corrosion rate measured before each NaCl progressive chloride addition

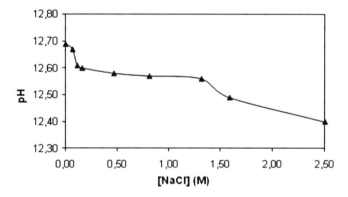

Fig. 4 pH monitoring during experiment with progressive chloride addition

the pH value remains fairly constant. From about 1.5M the pH again decreases until reaching a value of 12.4 at the end of the experiment.

3.2 Fixed chloride concentration

Longer time tests in a 2.5M and 0.5M NaCl solution with N_2 bubbling continuously, were also performed. *Figure 5* shows the mean corrosion potential of 6 steel bars immersed in 2.5M NaCl during 5 days and the mean corrosion potential of 3 bars immersed in 0.5M NaCl solution during one month. In both cases, the corrosion potential value decreases with time. The initial potential value was around -500mV for the steel immersed in 2.5M NaCl solution and -470mV for the steel in 0.5M, demonstrating more fluctuations than in the 2.5M NaCl solution.

The evolution of the corrosion rate is shown in *Figure 6*. The values of the 2.5M and 0.5M NaCl curves are an average of six and three steels, respectively. The same

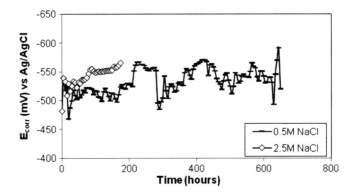

Fig. 5 Potential corrosion evolution for the steels immersed in 2.5M and 0.5M NaCl solutions

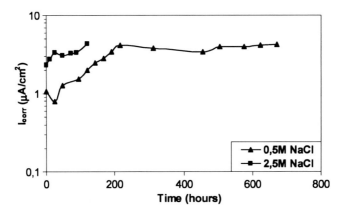

Fig. 6 Corrosion rate evolution for the steels immersed in 2.5M and 0.5M NaCl solutions

trend for the E_{corr} was found in both chloride concentrations. From the beginning of the test, the I_{corr} value is high: 2.33µA/cm^2 and 1.06 µA/cm^2 for 2.5M and 0.5M NaCl solution, respectively. I_{corr} values increased continuously with time of immersion until reaching values of 4.3µA/cm^2 for the 2.5M and 0.5M NaCl solution, respectively.

Applying Faraday's law it is possible to calculate the electrochemical weight loss [6]. The comparison of this electrochemical weight loss and the real gravimetric losses is shown in *Figure 7*. All values show a good enough accordance. Only the 2,5M solution presented weight losses not lying in the range of "two times accuracy" of the linear polarization technique but the mean is correct.

Despite the high values of corrosion rate measured, the visual observation at the end of the test in 2.5M only showed areas with corrosion stains and very small pits. *Figure 8* shows a picture of one of the steel tested for 5 days in 2.5M NaCl at free corrosion potential.

The oxides were more visible in the 0.5M NaCl solution test that lasted one month. Figure 9 shows some pictures of one of the steels tested in these conditions.

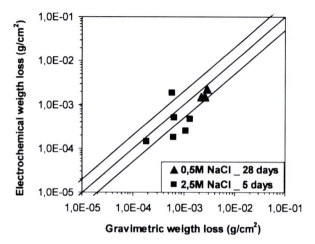

Fig. 7 Weight loss comparison for the steels immersed in 2.5M and 0.5M NaCl

Fig. 8 Appearance of the steel immersed during 5 days in 2.5M NaCl solution

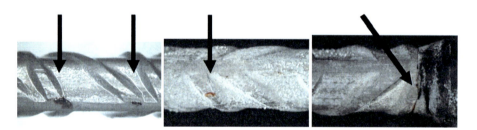

Fig. 9 Appearance of the steel immersed during 30 days in 0.5M NaCl solution

It can be seen that corrosion products were formed in a line along the bar which may be formed during the bar fabrication. Crevice corrosion is also observed at the edge of the tape used to delimit the working area.

In both cases (0.5M and 2.5M NaCl solutions) corrosion attack occurred only in areas with surface imperfections such as seams or rolled laps. Both the metallurgical defects and the interface with the tape used to delimit the working area, serve as sites for corrosion initiation. These imperfections seem to be the preferential sites for pit initiation and the border of the tape for crevice corrosion. It seems then that the high values of I_{corr} are truly indicative of active corrosion although the oxides are not visible to the eye until a certain amount is formed (more visible in the longer experiment of one month).

4 Discussion

The results clearly show that active corrosion develops in the presence of almost negligible oxygen concentration. Then the oxygen is not the main controlling factor for the development of corrosion and thus oxygen diffusion through the concrete cover cannot be the determining parameter of the corrosion kinetics.

For active corrosion the presence of chlorides is needed, however its proportion is not critical as it seems that the existence of defects in the steel surface acts simultaneously. That is, surface defects and chloride concentration are the main factors needed to develop and sustain active corrosion in spite of the quasi-absence of oxygen. The values of the corrosion rate can reach quite high values.

In Figure 10 several relationships are shown between the parameters studied in the test with progressive chloride addition, while in Figure 11 only one relationship in the test at fixed chloride concentration is shown. With respect to the relationship between the corrosion rate and the corrosion potential (Figure 10a and Figure 11) only a clear relationship for the fixed chloride amount can be demonstrated. In this case the change of potential for a decade of corrosion current is around 120 mV, precisely the expected Tafel slope for the oxygen reduction. However when the chloride is added progressively (trying to reproduce the real cases), there is not a clear direct relationship perhaps due to the fact that an equilibrium is not established or because the surface defects are more determining than the chloride content.

Regarding the relationship between I_{corr} and the Cl/OH ratio (Figure 10b) there is a direct relationship in the case of progressive addition which is very similar to that represented in Figure 3: the corrosion seems to activate for low concentrations and the rate increases with the chloride amount or the Cl/OH ratio. The trend is opposite when plotting I_{corr} versus the pH value. Also the corrosion potential follows the pH trend (Figure 10c). Whether the chloride increase or the pH decrease (Figure 10d), or both simultaneously, are responsible for the corrosion activation cannot be deduced as this will need further particular tests.

Concerning the places where pits and oxides nucleate, the surface of the steel reinforcement may present many imperfections as corrugated bars are used and its diameter is decreased during fabrication by hot-drawing leading to longitudinal bends which induce microscopic defects or small holes in which, differential aeration can be developed as in the tape border. Also the corrugation induces superficial areas with different roughness. All these factors seem to play an important role in the initiation and developing of active corrosion areas.

No explanation was found on how crevice corrosion was possible to develop, which is assumed to be induced by differential aeration, when almost no oxygen is present. Perhaps, the crevice develops with any difference in oxygen that is independent of the external oxygen content. It might also be that the potential is the parameter which plays the controlling role by inducing modifications in the passive layer and then allowing its dissolution. Although active corrosion has been detected in the present tests in chloride containing solutions, by continuous bubbling of N_2, many aspects of the corrosion process remain to be studied, in addition to the very basic steps of the depassivation phenomenon. Thus, it remains to clarify whether the

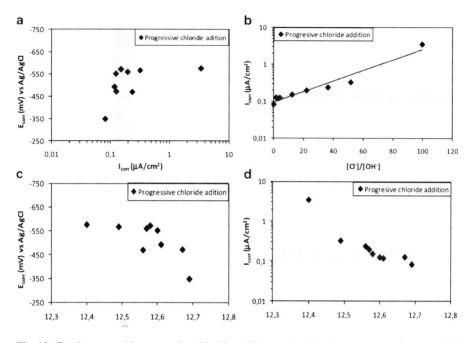

Fig. 10 For the tests with progressive chloride additions. **a)** relation between corrosion potential and corrosion rate, **b)** relation between corrosion rate and Cl/OH ratio, **c)** relation between corrosion potential and pH and **d)** corrosion rate and pH value

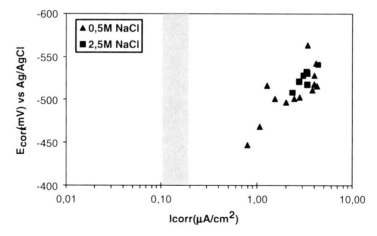

Fig. 11 Relation between corrosion potential and corrosion rate in the tests made with fixed chloride concentration

chloride concentration or the steel surface defects, is more important and how the pH value or the corrosion potential influences the depassivation onset. These questions do not should affect the statement that corrosion can develop in alkaline solutions in the presence of chlorides although oxygen is completely absent.

5 Conclusions

It is a general statement that the lack of oxygen in chloride solutions is a guarantee for the absence of corrosion. However, results reported above show opposite evidences. It is possible to have active corrosion due to chloride attack, even when the oxygen concentration in the electrolyte is negligible. The presence or absence of corrosion is influenced by the amount of chloride and the surface roughness.

The pits nucleate in the metallurgical defects and at the border of the tape used to delimit the working area. These imperfections seem to act as preferential sites for pit initiation and the border of the tape for crevice corrosion.

The oxides are not visible to the eye until a certain amount is formed and then the tests should last several days to visualize the active corrosion. It has been confirmed by the comparison between gravimetric and electrochemical losses that the high values of I_{corr} measured are truly indicative of active corrosion.

Aknowledgements The Authors thank the financial support from the Spanish Ministry of Education and Innovation through the project CONSOLIDER SEDUREC of the program INGENIO 2010.

References

[1] Pourbaix
[2] Izquierdo D., Alonso C., Andrade C., Castellote M. (2004). Potentiostatic determination of chloride threshold values for rebar depassivation - Experimental and statistical study, In: *Electrochimica Acta*, vol. 17–18, p. 2731–2739.
[3] Berke N.S. (1989). A review of corrosion inhibitors in concrete, In: *Materials Performance*, vol. 10, p. 41–44.
[4] Raupach, M. (1996). Investigations on the influence of oxygen on corrosion of steel in concrete - Part I, In: *Materials and Structures*, vol. 29, n. 187, p. 174–184.
[5] Raupach, M, Investigations on the influence of oxygen on corrosion of steel in concrete - Part II, In: *Materials and Structures*, vol. 29, n. 188, p. 226–232.
[6] Andrade C, González J.A. (1978). Quantitative measurement of corrosion rate of reinforcing steels embedded in concrete using polarization resistance measurements. In: *Werkstoffe und Korrosion*, vol. 29, p. 515.
[7] Rilem TC 154-EMC Recommendations. (2004). Test methods for on-site corrosion rate measurement of steel reinforcement in concrete by means of the polarization resistance method, In: *Materials and structures*, vol. 37. This is the very last row of this page. This is the very last row of this page.

Application of risk analysis in structural engineering – gas explosions

Ramon Hingorani and Peter Tanner

Abstract The present contribution describes the development of methods and tools intended for the practical application of explicit risk analysis in structural design. More specifically, it deals with the risk of personal injury caused by structural damage resulting from gas explosions, regarded as a significant threat to structural safety. Risk quantification entails the use of models able to quantify the uncertainties associated with the basic variables involved. A probabilistic model for the predominant action involved in explosions, explosion pressure, was deduced from test results available in the literature. This model was subsequently used to analyse the reliability level of a representative series of reinforced concrete beams against the implicit requirements laid down in building codes. The wide scatter observed is an indication of the lack of consistent calibration of current design rules. The findings may be used in future studies to deduce a consistent level of acceptable personal risk associated with gas explosion-induced structural collapse.

1 Introduction

Structural design codes must address the safety issue either implicitly or explicitly. When addressed explicitly, structure-related risks are quantified and compared to an acceptance criterion, which if based on rational principles, favours consistent structural safety decision-making. Inasmuch as the regulations presently in force establish only a general framework for explicitly addressing the safety issue, such methods have been virtually ignored in everyday practice to date. In order to bridge this gap, recent studies [1,2] have focused on developing methods and tools for use in the practical application of explicit risk analysis in structural design, with promising results.

R. Hingorani (✉) • P. Tanner
Instituto de Ciencias de la Construcción, Eduardo Torroja, Madrid
e-mail: hingorani@ietcc.csic.es; tannerp@ietcc.csic.es

The present contribution is a first step toward extending the scope of the work performed to date to design for accidental situations in which a building is exposed to exceptional conditions. More specifically, the paper deals with structure-related risks to persons due to gas explosions. The practical importance of this accidental situation, dealt with in the Eurocode [3], can be gleaned from the following statistic: a mean of approximately three moderate to severe gas explosions involving structural collapse with lethal consequences for building users occur yearly in Spain.

Risk acceptance is a complex issue governed not only by technical, but also ethical and especially political factors. A simple and logical procedure for establishing acceptable structure-related risks is to adopt the inherent risks set out in existing structural standards, which reflect general practice and are acceptable by definition. The definition of acceptable structure-related risk therefore depends on the degree of reliability implicitly required by such standards. For want of consistent calibration of the standardised rules, the reliability level implicitly required by the codes is unknown, however, and must therefore be determined. This entails prior development of probabilistic models able to quantify the uncertainties associated with the basic variables involved.

This paper introduces a probabilistic model for the predominant action involved, explosion pressure, deduced from test results available in the literature. It was used, along with other probabilistic models, to determine the reliability level implicitly required by structural design codes. A representative series of reinforced concrete beams exposed to the effects of gas explosions was analysed in the present study. The exercise explored dynamic effects such as energy dissipated due to plastic deformations of reinforcement and the strain rate dependent behaviour of materials. The results are presented in terms of the reliability index and the probability of structural failure.

2 Probabilistic models for variables involved in structural safety related to gas explosions

2.1 *General considerations*

Probabilistic models for the most prominent structural design variables that represent the uncertainty associated with the rules laid down in Spanish codes, compatible with the respective structural Eurocode specifications, were deduced in prior studies [4]. These models, suitable for practical application, are consistent with the models laid down in the Probabilistic Model Code [5]. They quantify uncertainty on the basis of statistical characterisation of the random variables: more precisely, each random variable involved is characterised by its statistical distribution, and its first two moments, mean value and standard deviation.

Although these probabilistic models were derived in the context of persistent design situations, some may also be adopted for reliability analysis in connection with gas explosions. Models designed for static load variables or referring to the geometric dimensions of structural members, for instance, require no modification for use in the context of accidental situations. Therefore, the present contribution focuses on the variables whose models differ from the models developed for persistent situations. One of these variables is the main load variable involved in a gas explosion, i.e., explosion pressure, and the effects of this load on structural members. In addition, the material models for concrete and steel had to be adapted to accommodate strain rate.

2.2 Gas explosion pressure

Gas pressure build-up in a vented chamber is quite complex because it depends on the interaction of numerous parameters [6], including:

- geometry of the room where the explosion occurs
- type, size and geometrical distribution of venting components
- composition of the gas cloud (fuel-air mixture)
- location of the point of ignition
- generation of turbulence around obstacles

Very little research has been conducted on predicting pressure in natural gas explosions. In the nineteen seventies, after a gas explosion induced the collapse of the Ronan-Point building in London, Dragosavic [7] conducted an extensive experimental study to deduce a method for predicting the explosion pressure that may occur in a standard dwelling. His systematic survey studied the effect of room geometry and different venting components on the pressure build-up inside a rectangular enclosure. His study led to a model that estimates the explosion pressure in terms of the relationship between the static failure pressure of venting components, p_{stat}, and the so-called venting coefficient, i.e. the ratio of venting component area, A_v, to the volume of the enclosure, V. The design value for explosion pressure, $p_{ex,d}$, proposed by Dragosavic was adopted by a number of structural design codes, Eurocode EN 1991-1-7 on accidental actions [3] among them. Figure 1 shows the design value of $p_{ex,d}$ versus the venting coefficient, A_v/V assuming a relatively high value for the static failure pressure of standard venting components (doors and windows), $p_{stat} = 7$ kN/m^2.

Dragosavic's findings [7] were analysed statistically in the present study to deduce a probabilistic model for gas explosion pressure usable in reliability calculations. An expression for the mean value, μ_{pex} was found with regression analysis, as illustrated in *Figure 1*. As the figure shows, the mean value found may exceed the design value for small venting coefficients A_v/V. This is mainly because Dragosavic reduced the explosion peak pressures to 80 % of their measured value before plotting the envelope used to deduce the design criteria. While a discussion of the assumptions that led to that reduction is beyond the scope of the present paper, it should nevertheless be borne in mind that situations in which the mean value is

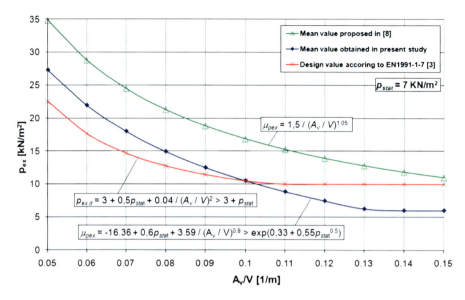

Fig. 1 Comparison of the explosion pressure design value [3], the mean value found in the present study and the mean value proposed in [8]

higher than the respective design value necessarily have an effect on the reliability calculated for the structural members analysed.

Moreover, *Figure 1* contains a model for predicting the mean value of the explosion pressure proposed by Schmidt [8], who conducted numerical simulations of natural gas explosions in dwellings based on a computational fluid dynamics model. The results show that for the p_{stat} value assumed, his suggested mean value is considerably higher across the entire range of A_v/V ratios than in the EN1991-1-7 [3] design criteria and the mean found in the present study, both based on Dragosavic's test results. Considering that these tests were conducted under the conservative assumption of a homogenously distributed, stoichiometric mix of gas, which seems highly unlikely under real circumstances [7], the mean value found with numerical and test results as shown in *Figure 1* may seem surprising. Schmidt's study [8], however, took other random parameters into consideration, whose effects would favor pressure generation. These parameters, which included the adverse location of the point of ignition with respect to the location of venting components and especially the generation of obstacle-induced turbulence, were not addressed in detail in the tests conducted by Dragosavic. This could explain the wide gap between Schmidt's mean value and the value found in the present study. However, since no tests that might confirm Schmidt's numerical results have been conducted, his model was not used here. The aforementioned model deduced by a regression analysis of Dragosavic's test results was used instead. A similar approach followed by [9] yielded a different model, which will not be discussed here.

The fitting of a statistical distribution to explosion peak pressure is a difficult exercise due to the short number of available test results in the literature. Further to the central limit theorem, Schmidt assumed a normal distribution [8], as did

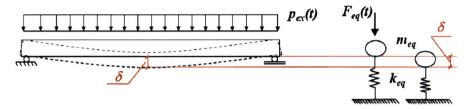

Fig. 2 Simplified dynamic analysis of concrete beams in an *SDOF* system

Leyendecker and Ellingwood [9] based on Dragosavic's test results. Nonetheless, fitting a normal distribution to the basic variable "explosion pressure" may lead, at least theoretically, to negative pressure values, which could not have been observed in the gas explosion tests analysed. A lognormal distribution may therefore be a more suitable approach and for that reason was adopted in the present reliability analysis.

Dragosavic conducted a maximum of seven tests under nominally identical conditions to study the scatter inherent in the generation of explosion pressure. Based on a statistical evaluation of the respective test results, Ellingwood and Leyendecker deduced a coefficient of variation of $CoV=0.36$, which they used in their Monte Carlo simulations to calibrate a partial coefficient for the gas explosion pressure load [9]. However, the present study showed that, at a 95 % confidence level, when the uncertainty stemming from the small number of test results available was taken into consideration, the coefficient of variation rose to approximately 0.7, which was the value adopted for the reliability analysis performed. For reasons of simplification and for want of more detailed data, this value was assumed to be constant across the entire range of possible values of A_v/V and p_{stat}. Vrouwenvelder and Leira, who consider the EN 1991-1-7 design pressure as a reasonable approximation for the pressure mean value, also indicate a CoV of around 0.7 [10].

The foregoing considerations are indicative of the high degree of uncertainty associated with the probabilistic model derived. A sensitivity analysis of the results obtained, varying probability distribution functions and, within reasonable limits, the magnitudes of their characteristic parameters, μ and CoV, merits future study.

2.3 Load effects on concrete beams in gas explosions

Explosion pressure acts like a dynamic load on the inner surfaces of the elements forming the compartment and is transferred from there to load carrying structural members such as slabs, beams and columns. To simplify the dynamic analysis, the beams analysed in the present study were assumed to be single degree of freedom (*SDOF*) systems consisting of a lumped mass connected by a spring to a fixed bearing (*Figure 2*). The mass and spring characteristics were properly defined to ensure they accurate reflected structural performance. Gas explosion pressure, assumed to act like a uniformly distributed load on the surface of the element, may be converted to an equivalent load $F_{eq}(t)$ to determine *SDOF* system loading. In doing so, the explosion pressure pulse characteristics, i.e., its time-dependent evaluation must be adequately represented. Dragosavic's tests [7] showed that the assumption of a triangular pulse seems to be reasonably suitable for gas explosions, a finding confirmed as well by the results of numerical simulations based on computational fluid dynamics [8]. The load pulses recorded in [7] acted for approximately $t_p = 0.2$ to 0.3 s on the closures of rather small room volumes of up to 36 m³, filled previously with a stoichiometric natural gas-oxygen mixture. Load pulse duration, t_p, may be assumed to rise with volume of the enclosure where the explosion occurs, as a rule [11]. On these grounds and assuming room volumes consistently smaller than 1 000 m³, t_p values from 0.2 s to 0.5 s were adopted depending on the span of the beam analysed. Large- span beams, present in larger volume enclosures, were consequently assumed to be exposed to load pulses of a longer duration.

A dynamic load factor (*DLF*) can be estimated by idealising the structural member (assuming an SDOF system) and the explosion pressure (adopting equivalent loading). This factor, introduced by Biggs [12], describes the ratio of dynamic to static structural response. A *DLF* accounting for elasto-plastic structural behaviour may be deduced as a function of the ratio of load pulse duration, t_p, to the oscillation period, T, of the structural member, and the so called displacement ductility factor, μ_δ, associated with the spring in the *SDOF* system, i.e., the ratio of the maximum displacement, δ_{max}, to yield displacement, δ_y [12]. In the present study, the displacement ductility factor was estimated from the curvature ductility factor of the beam mid-span sections analysed, μ_κ, i.e., the ratio of the ultimate curvature, κ_u, to the yield curvature, κ_y. Therefore, the simplified relationship expressed by Equation (1) was employed, taken from [13].

$$\mu_\delta = \frac{\mu_\kappa + 1}{2} \qquad (1)$$

Using μ_δ and t_p/T for an iteration-based numerical analysis, the ratio of the maximum dynamic bending moment to the respective statically applied peak pressure moment can be obtained by assuming that the mid-span section of the beam attains its ultimate deformability. This ratio, designated in the following discussion as DLF_{Mpex}, was introduced in the reliability analysis of the mid-span cross sections of the beams analysed in the present study. For analysing shear failure related reliability at support of the beams, a corresponding dynamic load factor DLF_{Vpex} may be obtained by assuming the beams to be in dynamic equilibrium and that the pattern of distribution of inertia across their span is the same as their assumed deflected shape [12].

Predicting load effects as described is obviously subject to uncertainties. Noting that models for the calculation of the loads on structural members due to explosions must be considered as highly uncertain, a coefficient of variation (*CoV*) related to model uncertainties of at least 0.3 was recommended in [14]. For the reliability analysis performed in the present study, this value was adopted for predicting both bending moment and shear forces due to explosion pressure. The distribution was assumed to be lognormal with a mean equal to one.

2.4 Material strength

It is widely known that both concrete compressive strength and the yield stress of the reinforcing steel are sensitive to strain rate. Gas explosions are regarded as a relatively slow type of impact loading [15] and consequently the increase of material strength is fairly small compared to high velocity impact scenarios. The reinforcement yield stress and concrete compressive strength are nonetheless believed to be substantially enhanced and have therefore been factored into the reliability analysis performed here. The approach followed was to represent material strength under high loading rates, or simple, dynamic material strength, $f_{m,dyn}$, as the product of the strength for the material associated with quasi-static loading, f_m, and a dynamic increase factor (DIF):

$$f_{m,dyn} = f_m \cdot DIF \tag{2}$$

Model Code 90 [16] contains a model adapted from the stochastic approach developed by Mihashi and Wittmann [17] to determine the effect of rising strain rates $\dot{\varepsilon}$ on concrete compressive strength. This frequently quoted model, shown as Equation (3) below, was used in the present study.

$$f_{c,dyn} = f_c \cdot \left(\frac{\dot{\varepsilon}}{3 \cdot 10^{-5}}\right)^{(1.026 \cdot \frac{1}{5+9 \cdot f_c/10})} \quad [\text{N/mm}^2], \dot{\varepsilon} \leq 30 [s^{-1}] \tag{3}$$

The increase in reinforcing steel yield stress under dynamic loading was computed in the present study with a model based on experimental research reported by Malvar and Crawford [18] as shown in Equation (4) below.

$$f_{ys,dyn} = f_{ys} \cdot \left(\frac{\dot{\varepsilon}}{10^{-4}}\right)^{(0.074 - 0.04 \cdot \frac{f_{ys}}{414})} \quad [\text{N/mm}^2], \dot{\varepsilon} \leq 225 [s^{-1}] \tag{4}$$

Estimation of the material strain rate $\dot{\varepsilon}$ is a pre-requisite to applying these models. For the sake of simplicity, a constant mean strain rate $\dot{\varepsilon}$ was derived for the present study, consisting of the ratio of the material strain due to explosion pressure, ε_{pex}, to the time required to attain this strain, $t_{\varepsilon,pex}$. Strain, ε_{pex}, can be obtained from the mid-span cross-section analysis and time $t_{\varepsilon,pex}$ can be computed from the *SDOF* calculations. For the beams analysed, these assumptions led to concrete and reinforcing steel strain rates of $\dot{\varepsilon} \approx 5 \cdot 10^{-4}$ to $\approx 10^{-2}$ s^{-1}. Given the simplifications adopted, these values are rough estimates only. Since Equations (3) and (4) are logarithmical, however, the prediction of $\dot{\varepsilon}$ need not be overly precise, as pointed out in [15]. Solving for the strain rates assumed here, the above expressions yielded dynamic concrete compressive strength values for reinforced concrete beams exposed to the effects of a gas explosion that were 5 to 20 % higher than static compressive strength. The dynamic reinforcing steel yield stress, in turn, was estimated to be approximately 5 to 12 % greater than the static value.

Equations (3) and (4) were used for reliability analysis of reinforced concrete beams. The strain rate, $\varepsilon = \dot{\varepsilon}$, was regarded as a deterministic variable for reasons of simplicity. The probabilistic models developed in former studies [4] were used to describe concrete compressive strength f_c and steel yield stress f_{ys} under quasi static loading. The scatter for dynamic material resistances, $f_{c,dyn}$ and $f_{ys,dyn}$, was assumed to be the same as in f_c and f_{ys}, respectively. As far as concrete is concerned, this assumption was theoretically validated by prior research [15].

3 Implicit reliability level of concrete beams exposed to gas explosions

3.1 Procedure

The procedure adopted to determine the level of reliability implicitly required by a consistent set of codes for structural design was outlined in prior papers [1,19]. In general terms, the aim consists of determining the reliability level of representative structural members designed strictly to a consistent set of codes, acknowledging that these members are by definition regarded as "safe enough" in the context of such codes. The procedure for determining the reliability level implicitly required by the codes is summarised below:

- selection of a representative series of structural members
- identification of the most representative failure mechanisms associated with the members selected and formulation of the respective limit state functions (LSF)
- establishment of the probabilistic models relating to the basic variables
- determination of the reliability levels for all the failure mechanisms identified and each strictly designed structural member
- statistical evaluation of the findings.

3.2 Representative series of concrete beams

The numerical exercise conducted in the present study referred to reinforced concrete beams in residential buildings only. The beams were assumed to be exposed to gas explosion pressure acting in the same direction as gravitational loads. In light of their widespread use in conventional buildings and the fact that their application in the model constitutes a conservative approach in terms of structural reliability, only statically determinate members were considered.

Varying the parameters with the greatest effect on design (such as use category, beam span, number of storeys, material strength and permanent loads) within reasonable ranges to cover the vast majority of cases encountered in practice yielded a

Application of risk analysis in structural engineering – gas explosions

representative series of 486 hypothetical but realistic concrete beams. The parameters and ranges are listed below:

- static failure pressure for standard venting components (windows and doors): p_{stat} = 2 to 7 kN/m²
- venting coefficients: A_v/V = 0.05 to 0.15 m⁻¹
- beam spans: l = 5 to 40 m
- characteristic concrete compressive strength: f_{ck} = 25 to 50 N/mm²
- characteristic yield stress for reinforcing steel: f_{yk} = 500 N/mm²
- variable load: q_k = 2 kN/m² (residential building use category [20])
- permanent loads: pk = 0.5 to 13,0 kN/m².

3.3 Failure mechanisms and limit state function

The most representative failure mechanisms for the beams selected were assumed to be mid-span bending failure and support section shear failure. For the latter one, failure of the compression strut was distinguished from tension tie failure. The respective limit state functions (LSF) for cross-sections were then deduced from the codes on basis of design, actions and resistance [20,21,22]. The impact of dynamic effects on beam reliability, i.e., the inertia effects taking account of elasto-plastic structural behaviour and the strain rate effects, was introduced in the LSF by applying a dynamic load factor, *DLF*, and a dynamic (material) increase factor, *DIF*, as described in the previous chapters. By way of example, the LSF for bending failure in a mid-span cross section is given by the following expression (5):

$$\xi_{R,M}\left(A_s \cdot f_{ys,dyn} \cdot d - 0.5\frac{(A_s \cdot f_{ys,dyn})^2}{b_v \cdot 0.85 \cdot f_{c,dyn}}\right) - \xi_{E,Mpex} \cdot DLF_{Mpex} \cdot M_{pex}$$
$$-\xi_{E,M} \cdot (M_c + M_p + \psi_1 \cdot M_q) = 0$$

$\xi_{R,M}$:   model uncertainty coefficient for determining resisting bending moments

$\xi_{E,M}$:   model uncertainty coefficient for determining bending moments due to static loads

$\xi_{E,Mpex}$:   model uncertainty coefficient for determining bending moments due to explosion pressure

M_c, M_p, M_q, M_{pex} :   bending moments due to self weight, g_c, permanent loads, g_p, variable loads, q and statically acting peak explosion pressure, p_{ex}, respectively

ψ_1 :   load combination coefficient to be taken into account in accidental situations [21]

A_s :   cross-sectional area of longitudinal reinforcement

b_v, d : cross section width and inner lever arm, respectively
$f_{c,dyn}, f_{ys,dyn}$: concrete compressive strength (3) and reinforcing steel yield stress (4) when exposed to loads of short duration
DLF_{Mpex} : dynamic load factor for converting the statically applied peak pressure corresponding moment into the acting dynamic bending moment taking account of inertia and energy dissipation effects in the structural members when exposed to loads of short duration

3.4 Strict Design

After the limit state functions were defined, the mid-span and support sections of each reinforced concrete beam selected was designed strictly to the specifications of the consistent set of codes for which the implicit level of reliability was to be determined [20,21,22]. Strict design means that structural member performance complied exactly with the safety requirements laid down in the codes in question. The design material resistances and the design load effects were determined according to the accidental situation design rules. The design accidental action considered, i.e., the explosion pressure, was taken from the EN 1991-1-7 [3] model described earlier, as well contained in [20].

3.5 Reliability analysis

Finally, the reliability level was determined in terms of the probability of failure, p_f, and the reliability index, β, for each representative failure mechanism in all the strictly designed beams. The results were obtained using the FOSM algorithm [23] set out in code VaP 3.0 [24]. The probabilistic models used for this purpose were the models for the variables representing the state of uncertainty associated with the rules laid down in the codes analysed, as described in the previous chapter. In keeping with standard practice, the reliability level was referred to a building service life of 50 years.

The probabilities of failure obtained for the three failure modes in all 486 beams analysed are shown in *Figure 3*, versus the ratio of variable loads to total loads, v. The results shown must be interpreted as conditional failure probabilities, i.e. the probabilities of structural failure given a gas-induced explosion. This conditional probability is designated as $p_{f,ex}$ in the following discussion. The first finding was that for all the failure mechanisms analysed, the conditional failure probability, $p_{f,ex}$, rose with increasing v. This confirms the heavy impact of variable actions on the reliability level of the beams analysed. An analysis of the sensitivity coefficients α [23] identified explosion pressure as the variable with the greatest effect on the

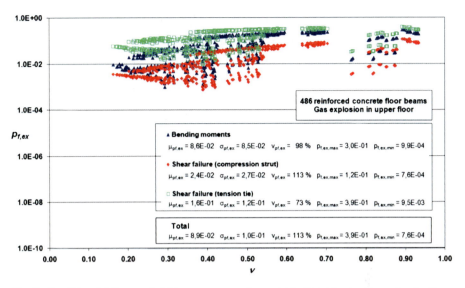

Fig. 3 Conditional failure probability, $p_{f,ex}$, versus the variable / total load ratio, v, for the three failure mechanisms analysed in 486 reinforced concrete floor beams (reference period: $T_{ref} = 50$ years)

reliability of the members studied. This may be attributed to the high degree of uncertainty associated with the generation of pressure, represented in the reliability analysis by a relatively high coefficient of variation, as discussed earlier.

As shown in *Figure 3*, the lowest probability of failure was associated with shear failure in the compression strut. The minimum value of $p_{f,min,ex} = 7{,}6 \cdot 10^{-4}$ was obtained for a beam with a variable to total load ratio of $v = 0{,}39$, while the highest probability, $p_{f,max,ex} = 3{,}9 \cdot 10^{-1}$, was found for tension controlled shear failure, with a v of 0,92.

An analysis of the mean probability of failure revealed that the value for compression-controlled shear failure, $\mu_{pf,ex} = 2{,}4 \cdot 10^{-2}$, was four times smaller than the value associated with mid-span bending failure and seven times smaller than the value for tensile failure of the shear reinforcement. The mean value for all the conditional probabilities of failure, $\mu_{pf,ex} = 8{,}9 \cdot 10^{-2}$, was an indication that structural failure would be expected in about 9 % of the reinforced concrete beams analysed after exposure to a gas explosion.

Scatter was of the same order of magnitude in all the failure mechanisms analyzed. The coefficients of variation fluctuated from $v_{pf,ex} = 73$ % in the tension governed shear failure to $v_{pf,ex} = 113$ % for shear failure due to excessive stress on the compression strut. These findings show that the reliability level for the concrete floor beams analysed, designed strictly to the rules of a consistent set of structural design codes, is affected by a wide dispersion

As noted earlier, the results shown are conditional failure probabilities, i.e. the probabilities of structural failure assuming a prior gas explosion. Since such explosions are non-persistent or extraordinary events, however, the likelihood of their

Table 1 Reliability index, β, and probability of failure, p_f, in reinforced concrete beams in gas explosion and persistent design situations (reference period: 50 years)

Design situation	No. of beams analysed	No. of failure mechanisms	No. of results	Reliability index, β μ_β	v_β	Probability of failure, p_f μ_{pf}	v_{pf}
Gas explosion	486	3	1458	4,1	0,09	$4,5 \cdot 10^{-5}$	1,13
Persistent	450	3	1350	4,2	0,06	$2,4 \cdot 10^{-5}$	1,04

occurrence must also be taken into account. Pursuant to a frequently applied approach [8,9], the probability of failure of a given structural member, p_f, induced by gas explosion within a given reference period, T_{ref}, may be estimated as the product of the likelihood of a gas explosion, $p(ex)$, times the probability of failure of the structural member in the event of such explosion, $p_{f.ex}$.

$$p_f = p(ex) \cdot p_{f.ex} \qquad (6)$$

The probability of occurrence of an exceptional event such as a gas explosion within a given reference period T_{ref} can be found with a Poisson distribution approach [8]. As Equation (7) shows, such calculations require information on annual mean occurrence of gas explosions per dwelling with a natural gas connection, λ_{ex}. Despite the absence of such data for Spain, the good correlation among the findings reported by researchers in Germany [8], the United States [9] and the Netherlands [10], allows for an assumption of $\lambda_{ex} \approx 10^{-5}$ to 10^{-6} for this parameter, from which the first mentioned value was employed for the purpose of the present study. Consequently, the likelihood of gas explosion in a 50-year period found with Equation (7) was $p(ex) \approx 5 \cdot 10^{-4}$.

$$p(ex) = P(x > 0 | \lambda_{ex}, T) = 1 - e^{-\lambda_{ex} \cdot T} \approx \lambda_{ex} \cdot T \qquad (7)$$

Entering these data into Equation (6) yields the probability of failure p_f. The respective reliability index, β, may then be deduced by applying the inverse standard normal distribution, Φ^{-1}.

$$\beta = -\Phi^{-1}(p_f) \qquad (8)$$

The p_f and β mean values and coefficients of variation for all the results obtained are given in *Table 1*, which also shows the results of a previous study [19] on the reliability of similar beams in persistent design situations for comparison.

The conclusion to be drawn from the above comparison is that the reliability level of reinforced concrete beams implicitly required by design codes is nearly the same for these two design situations. Moreover, in both cases the mean value of the reliability index, β, was higher than the value implicitly required in the code: $\beta = 3.8$. If the reliability index is assumed to follow a normal distribution, however, about 19 % of the reinforced concrete beams designed strictly to the rules for gas explosion failed to meet the $\beta = 3.8$ requirement, whereas in beams designed for persistent situations only 6 % of the beams were non-conforming.

Finally, the coefficients of variation for the probability of failure given in Table 1 reveal that reliability level associated with reinforced concrete floor beams in both gas explosion and persistent design situations is widely scattered. This may be attributed primarily to the lack of consistent calibration of the design rules.

4 Conclusions

Recent studies [1,2] have focused on developing methods and tools for use in the practical application of explicit risk analysis in structural design, with promising results. The present contribution constitutes a first step toward extending the scope of the work performed to date to design for accidental situations in which a building is exposed to exceptional conditions. More specifically, in light of the practical importance of the issue, the effect of gas explosions on residential buildings is explored.

First, a probabilistic model was introduced for the dominant action involved, explosion pressure, deduced from test results available in the literature. Further to design recommendations laid down in building codes, the mean value was expressed in terms of the venting coefficient A_v/V, and the static release pressure of the venting components, p_{stat}. A lognormal distribution was adopted and, in view of the small number of available test results, a constant coefficient of variation of 0.7 was estimated.

Together with other so-called implicit probabilistic models, which suitably reflect the uncertainty associated with the rules laid down in Spanish structural codes, the explosion pressure model was used to determine the reliability implicitly required by these codes. A representative series of strictly designed reinforced concrete beams exposed to gas explosion pressure acting in the same direction as gravitational loads was analysed. This analysis included a study of dynamic effects such as the energy dissipated as a result of the plastic behaviour of flexural reinforcement and the strain rate related sensitivity of materials. The results, shown in terms of the probability of structural failure, were found to be widely scattered, an indication of the lack of consistent calibration of current design rules. The mean value of the conditional failure probability of the beams analysed (in which the condition was a prior gas explosion in a 50-year reference period) was a troubling, $\mu_{pf,ex} = 8.9 \cdot 10^{-2}$. Factoring the likelihood of a gas explosion into the model, however, lowered the mean failure probability of the beams analysed to $\mu_{pf} = 4.5 \cdot 10^{-5}$. This value was found to be very similar to the value for a series of like beams under persistent design situations analysed in former studies. An analysis of the results for reliability index β revealed that in both gas explosion and persistent design situations, the mean value of β implicitly met the code requirement for that parameter (3.8). That notwithstanding, about 19 % of the reinforced concrete beams designed strictly to the gas explosion-related design failed to fulfil the $\beta = 3.8$ requirement, providing further evidence of a need for consistent calibration of the design rules. The results obtained here can be used in future studies geared to deducing consistent acceptance criteria for structure-related risks to persons in the wake of gas explosions.

References

[1] Tanner, P. and Hingorani, R. (2010), *Development of Risk-based Requirements for Structural Safety*, IABSE-fib Conference 2010, Codes in Structural Engineering, Dubrovnik

[2] Tanner, P. (2006), *Safety and reliability in structural engineering. Implicit vs. explicit approaches*, Concrete Solutions, 2nd International Conference on Concrete Repair, St Malo, ISBN 1-86081-915-X

[3] EN 1991-1-7 (2006), *Eurocode 1 – Actions on structures – Part 1-7: General actions – Accidental actions*, European Committee for Staandardization, Brussels

[4] Tanner, P. and Lara, C. (2005), *Simplones y conservadores. Modelos probabilistas para la evaluación de estructuras existentes de hormigón*, III Congreso de ACHE, Comunicaciones, vol. 3, Madrid. ISBN 84-89670-53-6.

[5] JCSS (2001), *Probabilistic Model Code*, Joint Committee on Structural Safety, www.jcss.ethz.ch

[6] Bjerketvedt, D., Bakke, J.R. and Van Wingerden, K. (1997), *Gas Explosion Handbook*, Journal of hazardous materials 52, pp. 1–150

[7] Dragosavic, M. (1973), *Structural measures against natural-gas explosions in high rise blocks of flats*, Heron, n° 4, pp. 5–51

[8] Schmidt, H. (2003), *Versagenswahrscheinlichkeiten unbewehrter Wand-Decken-Verbindungen bei Gasexplosionen im Fertigteilbau*, Dissertation, Technische Universität Darmstadt

[9] Leyendecker, E.V. and Ellingwood, B.R. (1977), *Design methods for reducing the risk of progressive collapse in buildings*, National Bureau of Standards, Building Science Series 98, U.S. Government Printing Office, Washington, D.C.

[10] Vrouwenvelder, T. and Leira, B. (2009), *Probabilistic modelling of internal gas explosions*, Joint Workshop of COST Actions TU0601 and E55, Ljubljana, Slovenia, www.cost-tu0601.ethz.ch

[11] Beshara F.B.A. (1994), *Modelling of Blast Loading on Aboveground Structures-II. Internal Blast and Ground Shock*, Computers & Structures 51 (5), pp. 597–606

[12] Biggs, J.M. (1964), *Introduction to Structural Dynamics*, McGraw-Hill, ISBN 07-005255-7.

[13] EN 1998-1-1 (2004), *Eurocode 8: Design of structures for earthquake resistance – Part 1: General rules, seismic actions and rules for buildings*, European Committee for Staandardization, Brussels

[14] Vrouwenvelder, T. and Stiefel, U. (2008), *Modelling of Internal Explosions*, Basic note for JCSS, COST TU0601 and TC250/SC1

[15] CEB (1988), *Concrete Structures under Impact and Impulsive Loading*, Bulletin d'information No. 187, Comité Euro-International du Béton, Lausanne, Switzerland

[16] CEB-FIB Model Code (1990), Comité Euro-International du Béton, Redwood Books, Trowbridge, Wiltshire, UK

[17] Mihashi, H. and Wittmann, F.H. (1980), *Stochastic approach to study the influence of rate of loading on strength of concrete*, Heron 25, n° 3

[18] Malvar, L.J. and Crawford, J.E. (1998), *Dynamic increase factors for Steel Reinforcing bars*, Twenty-eighth DDESB Seminar, Orlando, Fl

[19] Tanner, P., Lara, C., and Hingorani, R. (2007), *Seguridad estructural. Una lucha con incertidumbres*, Hormigón y Acero n° 245, Madrid, ISSN: 0439–5689

[20] CTE DB-SE-AE (Borrador) (2003), *Codigo Técnico de la Edificación, Documento Básico: Acciones en la Edificación*, Ministerio de Fomento, Madrid

[21] CTE DB-SE (Borrador) (2003), *Codigo Técnico de la Edificación, Documento Básico: Seguridad Estructural*, Ministerio de Fomento, Madrid

[22] EHE (1998), *Instrucción de Hormigón Estructural*, Ministerio de Fomento, Madrid, ISBN 84-498-0390-X.

[23] Hasofer, M., and Lind, N.C. (1974), *Exact and Invariant Second-Moment Code Format*, Journal of Engineering Mechanics, ASCE, Vol.100, EM1, pp.111–121

[24] VaP 3.0 (2010), *Variables Processor*, PSP, Feldkirchen, Austria

Hydrogen induced changes in structural properties of iron: Ab initio calculations

Alejandro Castedo, Javier Sanchez, José Fullea, Carmen Andrade, and Pedro Luis de Andres

Abstract We have used Ab initio calculations to study the structural changes produced by the inclusion of light impurities into pure bcc iron. In order to obtain a clear picture of the mechanics of the phase changes Bain's pathway was studied in detail for pure iron. The position that hydrogen atoms tend to occupy at high densities favours octahedral sites inside the bcc matrix, producing an internal stress field that suggests a deformation that matches the prediction of martensitic transformation predicted by Bain's pathway. We have used Density Functional Theory in order to optimize the structures studied, obtaining the enthalpy of the configuration as a function of c/a, allowing a better understanding of the dynamics of the process of phase changes.

1 Introduction

The study of interstitial impurities inside metals is of great importance when trying to understand the mechanical properties of these materials. Hydrogen embrittlement in particular is one of the processes that is driven by the action of these impurities

---

A. Castedo
Instituto de Ciencias de la Construccion Eduardo Torroja Consejo Superior de investigaciones Cientificas, Madrid, Spain

Instituto de Ciencias de Materiales de Madrid Consejo Superior de investigaciones Cientificas, Madrid, Spain

J. Sanchez (✉) · J. Fullea · C. Andrade
Instituto de Ciencias de la Construccion Eduardo Torroja Consejo Superior de investigaciones Cientificas, Madrid, Spain
e-mail: javiersm@ietcc.csic.es

P.L. de Andres
Instituto de Ciencias de Materiales de Madrid Consejo Superior de investigaciones Cientificas, Madrid, Spain

Fig. 1 Bcc, fcc and hcp structures

inside high-strength steels, composed primarily by bcc-iron (also referred as α-iron). The first discussion is whether hydrogen atoms "prefer" Octahedral sites (O-sites) over Tetrahedral sites (T-sites) inside ferromagnetic bcc iron. The energy difference between these two high-symmetry sites is small and the controversy about which is the actual absorption site has been there for a long time. Previous studies carried out in the past have shown that indeed both sites can be occupied, with the important difference that hydrogen is more likely to occupy T-sites when the system is encountered at a low density phase, whereas O-sites are preferred at large concentrations of hydrogen [1–3]. Experimental evidence tells us that under normal conditions the first case will be found but several recent works have suggested evidence of high concentrations of hydrogen. As predicted in past papers the inclusion of hydrogen in these O-sites produces an internal stress that results in a tetragonal distortion of the lattice.

In this work we focus our research on the possible structural changes of the lattice that might be produced by this tetragonal distortion mentioned before. There are several known phases of iron reported in literature. Experimental work shows three main phases: bcc iron (α-iron), fcc iron (γ-iron) and hcp iron (ε-iron). These three phases are shown in Fig. 1. Bain's pathway [4] offers an elegant and easy to understand mechanism for the martensitic transformation between α-iron and γ-iron. If we name a, b and c the axes of our unit cell (UC) we can explain this mechanism in terms of the change in the a/c ratio (a=b throughout the whole process). The tetragonal distortion consists of a contraction upon two of the cubic axes (a and b) and a continuous expansion along the third one (c). When the value $\sqrt{2}$ is reached the configuration of the system now matches that of the fcc iron. Our calculations go up to a ratio c/a = 1.6 where the hcp stable phase of iron has been found, although the mechanism that transforms fcc into hcp differs from that of Bain, since we no longer find the UC in the bct regime (one angle of the UC deforms to reach 60°) and it can be explained by a shear and a shuffle. This is the actual unit cell used in the calculations, a bct cell that consists of 2 atoms (or 3 in the case when we add hydrogen).

Bain's path is not the only transformation suggested in literature to explain this phase changes but it is preferable for a number of reasons: (i) it retains the highest possible crystal symmetry throughout the process, (ii) it is simple, and (iii) it contains the lattice deformation expected for hydrogen occupation of (O-site) interstitials.

2 Method and Theory

Ab initio calculations have been carried out in the framework of DFT [5] and pseudo-potentials theory [6]. Actual calculations have been performed with the CASTEP code [7]. Density Functional Theory (DFT) is a powerful tool that allows us to study and calculate the ground state of structures as long as we use correct and accurate descriptions for the exchange-correlation energy of the electrons of the system. Our calculations focus on geometry optimization procedures. We position the atoms in certain configurations allowing free movement of the atoms, so they can "look for" the lowest-energy configuration possible. One has to be careful when considering this minimums, since not always the lowest (global minimum) is found. Stable phases of different nature are encounter when performing this calculations, and the transitions between different phases have to be interpreted carefully. The Born-Oppenheimer approximation is used; ions are considered classical objects moving under the forces created by electrons obeying the Schrodinger equation. A plane-wave basis is used to expand electronic wave functions, and the cut-off energy is set to 375 eV. Ultrasoft pseudo-potentials are used to describe Fe and H, and the generalized gradients approximation for the exchange and correlation potential due to Perdew, Burke and Ernzernhof has been chosen [8]. Spin-polarized bands are considered to account for magnetism. The accuracy of our calculations is determined by the quality of the pseudo-potential, the cut-off energy, and the density of the k-points mesh used in the irreducible part of the Brillouin zone (a Monkhorst-Pack mesh of 10x10x10 [9]). The choice of these values for our calculations is related to previous studies where conditions for good convergence on relevant properties of iron such as the lattice parameter, magnetization, or bulk modulus have been assessed [2]. In order to establish when a configuration is truly stable we use certain convergence thresholds: variations in the total energy $\leq$ 10e-5 eV, maximum residual forces $\leq$ 0.001 eV/Å, and a maximum change in any atom position $\leq$ 0.001Å. The parameters of the UC (angles and distances, a, b and c) have been optimized in order to minimize the stress for each crystal system considered (cubic, tetragonal and hexagonal).

3 Results and Discussion

Our calculations are performed for the following system conditions, T= 0 K and P = 0 GPa. The global minimum for pure iron (without impurities) is a bcc phase corresponding to a ferromagnetic state, c/a = 1 and all angles 90°. The next minimum found corresponds to a minimum close to a fcc configuration, angles 90° and c/a = 1.51 and it is an antiferromagnetic (Type I; AF-I) state. The ideal ratio for a fcc phase is $\sqrt{2} \approx 1.41$ which can be found by our calculations if we consider

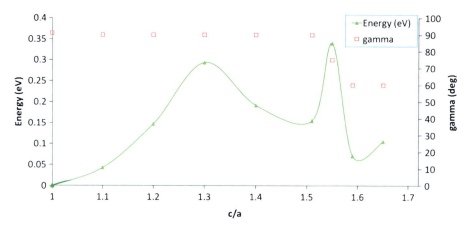

Fig. 2 Total energy along Bain's path for pure iron parameterized by the ratio c/a. The values of the angle γ is indicated above

NM (non magnetic) states. Once we consider the possibility of AF states we realized that an AF-I state would decrease the energy of this specific local minimum, making it the most possible candidate for the local minimum. Further away in the c/a ratio we find another local minimum with an hexagonal configuration (hcp), with γ = 60° and c/a ≈ 1.6 that matches the prediction for the ideal ratio of hcp structures as well as experimental examples of this phase. Magnetism is important and in the present work three possible magnetic ordering states have been considered for each minimum; Non-magnetic (NM), Ferromagnetic (FM) and Antiferromagnetic (AF). Only the lowest energy states are displayed in the figures that accompany this article.

In the case in which we add hydrogen the picture is qualitatively different. First we notice that hydrogen produces the so-called tetragonal distortion of the UC and bcc is no longer a minimum. The system is no longer at a stable configuration and the effect of this internal stress is to drive the system along Bain's path towards a fcc phase first, and subsequently towards an hcp phase that can be reached by a shear and a shuffle of our bct unit cell. This last configuration is the global minimum. We observed that the position of hydrogen slowly varies throughout the process choosing always an O-site, including the hexagonal phase only that this time it is harder to visualize. This behaviour agrees with experimental findings where hydrogen always occupies O-sites in hexagonal-close-packed structures, with no experimental evidence of occupancy of T-sites [10]. These experimental findings are also a good evidence of high density hydrogen systems.

The interpretation of the results obtained by our calculations has been summarized in Fig. 2. and Fig. 3. Table 1 gives actual numerical values compared to those available from experimental works.

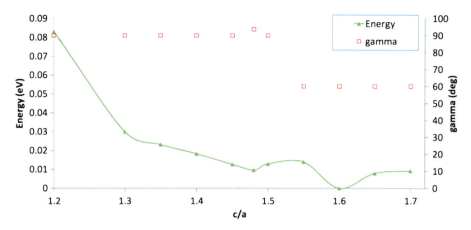

Fig. 3 Same as Figure 2 but for the system with hydrogen

Table 1 Comparison of theoretical and experimental parameters. Distances are given in Å, spin in μB and volume in Å³

	a(Å)	c/a	μB	vol (Å³)
Clean Fe				
BCC-Teor	2,82	1,00	2,20	22,33
BCC-Exp[11]	2,87	1,00	2,20	2,80
FCT-Teor	2,43	1,50	0,00	20,27
FCC-Exp[11]	2,55	1,42	0,00	-
HCP-Teor	2,45	1,58	0,00	22,16
HCP-Exp[10]	2,58	1,62	0,00	-
Fe+H				
HCP-Teor	2,52	1,60	0,00	22,13

Bibliography

1. Jiang, D.E. and E.A. Carter, *Diffusion of interstitial hydrogen into and through bcc Fe from first principles*. Physical Review B, 2004. **70**(Copyright (C) 2009 The American Physical Society): p. 064102.
2. Sanchez, J., et al., *Hydrogen in alpha-iron: Stress and diffusion*. Physical Review B (Condensed Matter and Materials Physics), 2008. **78**(1): p. 014113.
3. Sanchez, J., et al., *Ab initio molecular dynamics simulation of hydrogen diffusion in alpha-iron*. Physical Review B, 2010. **81**(13): p. 132102.
4. Bain, E.C., *The nature of martensite*. Transactions of the American Institute of Mining and Metallurgical Engineers, 1924. **70**: p. 25–46.
5. Kohn, W. and L.J. Sham, *SELF-CONSISTENT EQUATIONS INCLUDING EXCHANGE AND CORRELATION EFFECTS*. Physical Review, 1965. **140**(4A): p. 1133–&.
6. Vanderbilt, D., *SOFT SELF-CONSISTENT PSEUDOPOTENTIALS IN A GENERALIZED EIGENVALUE FORMALISM*. Physical Review B, 1990. **41**(11): p. 7892–7895.

7. Clark, S.J., et al., *First principles methods using* CASTEP. Zeitschrift Fur Kristallographie, 2005. **220**(5–6): p. 567–570.
8. Perdew, J.P., K. Burke, and M. Ernzerhof, *Generalized gradient approximation made simple.* Physical Review Letters, 1996. **77**(18): p. 3865–3868.
9. Monkhorst, H.J. and J.D. Pack, *SPECIAL POINTS FOR BRILLOUIN-ZONE INTEGRATIONS.* Physical Review B, 1976. **13**(12): p. 5188–5192.
10. Antonov, V.E., et al., *Neutron diffraction investigation of the dhcp and hcp iron hydrides and deuterides.* Journal of Alloys and Compounds, 1998. **264**(1–2): p. 214–222.
11. Oriani, R.A., *MECHANISTIC THEORY OF HYDROGEN EMBRITTLEMENT OF STEELS.* Berichte Der Bunsen-Gesellschaft Fur Physikalische Chemie, 1972. **76**(8): p. 848–857.

Corrosion initiation and propagation in cracked concrete – a literature review

José Pacheco and Rob Polder

Abstract The major degradation mechanism in civil engineering concrete structures is corrosion of reinforcement due to chloride penetration. Corrosion reduces serviceability and safety due to cracking and spalling of concrete and loss of steel cross section. Recently, service life design has moved from prescriptive performance based. The current approach aims at postponing initiation of corrosion until the end of the required service life with a predetermined reliability, based on simplified modelling of transport in uncracked concrete and testing of laboratory samples for chloride diffusion. Real structures under service load contain cracks and execution defects. Cracks are fast transport routes for chloride, but the effect is mitigated by poorly known mechanisms such as self-healing and crack blocking. Current models do not cover the effect of cracks, voids and compaction defects in concrete on chloride transport and corrosion initiation, rendering them less robust than desired. A project is carried out aimed at modelling the influence of cracks on the initiation and propagation of reinforcement corrosion. As the first phase, a literature review was made, which is reported in this paper.

J. Pacheco (✉)
Technical University Delft, Faculty Civil Engineering and Geosciences,
Section Materials and Environment, Delft, The Netherlands
e-mail: J.PachecoFarias@tudelft.nl

R. Polder
Technical University Delft, Faculty Civil Engineering and Geosciences,
Section Materials and Environment, Delft, The Netherlands
TNO Building Engineering & Civil Engineering, Delft, The Netherlands

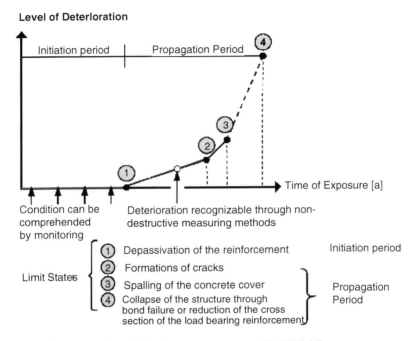

Fig. 1 Corrosion process for reinforced concrete structures (fib 2006) [1]

1 Introduction

Reinforcement corrosion is regarded as the major threat affecting serviceability and safety of reinforced concrete structures. In corrosion terms, the total life span of a structure involves two periods of time: initiation and propagation. A schematic representation regarding these periods is shown in Figure 1 [1]. The initiation period is considered as the time required until the passive layer on the reinforcement is removed by the action of either chlorides or reduction of alkalinity in the pore solution by carbonation. As soon as the chloride content is high enough or carbonation has reached the surface of the steel bar, the passive layer is dissolved and its protective function is lost. During the propagation period the steel reinforcement corrodes, reducing the bar section and therefore, the safety of the structure. Corrosion products are formed on the bar surface, producing tensile stresses leading to cracking, spalling and finally collapse.

Structural regulations focused on durability control allowed crack widths based on exposure conditions. Eurocode 2 and ACI 224 permits 0.3 mm and 0.15 mm, respectively, as the maximum value for marine environment. However, the degree of protection conferred by controlling the crack width depends on more factors that are not considered in the codes. For example, concrete composition, cover depth, compaction, environmental conditions and curing have significant effects on transport properties [2]. A study carried out by Otieno et al [3], concluded that it is not possible to determine a universal crack width. The highest tolerable crack width is a function of the concrete properties; therefore, a distribution of crack widths would be more adequate. Nevertheless, the effect of cracks on the corrosion mechanisms is still under debate. Some researchers consider that cracks have an effect only on the initiation period [4-8]. Cracks provide fast routes for chloride ingress, leading to quick depassivation of steel. In the propagation phase, other factors are believed to be more relevant: cover depth, concrete quality and environmental action. Also, macro-cell behaviour is considered to affect the performance of steel intersected by cracks. Other studies consider that the effect of cracks can be operating in both periods [9-11]. Not only the crack width is relevant during the initiation phase, but also the crack spacing determines the rate of deterioration. According to some views, even the crack orientation has a considerable effect on the corrosion process. Transverse cracks (i.e. bending cracks) intersect the steel reinforcement in a singular or several locations. At such points, local depassivation occurs. On the other hand, longitudinal cracks run along the orientation of the steel bars. This type of crack allows much higher chloride access to the surface of the steel and thus higher corrosion rates [12].

At the present time, service life design considers the period of serviceability until the end of the initiation period. In this sense, the service life of cracked structures could be significantly reduced. However, data gathered from the field shows that the performance of deteriorated structures does not follow this behaviour. It is clear that additional factors, which are still under investigation, are influencing the performance of cracked structures. In this sense, an extended study of the most important parameters involved in the corrosion process in cracked concrete would be valuable to determine if the current building codes are too restrictive or permissive.

2 Corrosion initiation in cracked concrete

Chloride ingress takes place through the concrete cover. Reinforced concrete in loaded structures by definition contains cracks, as the tensile capacity of the steel only works when the concrete is cracked. In uncracked concrete, a protective film is produced on the steel surface preventing corrosion (passivation). Under chloride attack, the passive film is destroyed and the steel starts to corrode actively. The time period until depassivation is reached is termed the initiation phase. Cracks allow chlorides to penetrate rapidly even through fine cracks, inducing corrosion initiation, as shown in Figure 2 [13]. When steel remains unprotected, this passive film is never produced. This situation is similar to bare steel in contact with a chloride-rich solution. Steel reinforcement in the crack opening would have a similar behaviour as unprotected, reducing the initiation period (or even nullifying it) when compared to steel in concrete. The causes are differences in the transport properties of uncracked and cracked concrete in terms of chloride transport and critical chloride content. The relation between the crack width and the initiation period in cracked concrete still needs further research.

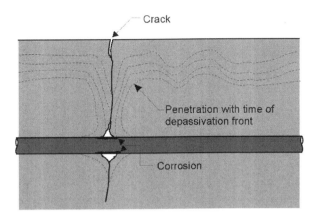

Fig. 2 Chloride ingress through cracks [13]

2.1 Chloride transport

Chlorides are transported into concrete by advection, capillary suction, diffusion and migration (if electrical fields are present). The rate of chloride ingress through cracks is of relevance for determining the length of the initiation period. Concrete properties that influence the transport of chlorides include cement type, water-binder ratio, air content and the chemical composition of the pore solution. Results obtained by Win et al [14] show a linear relationship between w/b ratio and chloride concentration for a determined crack width in concrete specimens. Also, the effect of capillary suction should not be disregarded. The influence of transverse cracks on chloride diffusion was studied by Djerbi et al [15]. It was found that the diffusion coefficient of cracked concrete (D_{cr}) increased with an increasing crack width. The D_{cr} was found to be constant when the crack width was higher than 80 μm. Boulfiza et al [16] studied a model considering the effect of combined action of different transport mechanisms. Results showed that the crack width and spacing had significant relevance under certain conditions that are not effectively modelled by Fick's Second Law. Therefore, the transport processes occurring inside a crack are more complex, requiring additional experimentation for determining a threshold value of crack width. Konin et al [17] found that the composition of concrete specimens with microcracks was critical for chloride ingress. The combination of high strength concrete and low w/b ratio reduced the chloride content noticeably. Also, a relationship between the apparent diffusion coefficient (D_a) and concrete strength was noticed. Finally, a higher crack density increased significantly the permeability of concrete, causing higher amounts of penetrating chlorides. The crack width/cover depth ratio was addressed by Gowripalan et al [18] as a suitable parameter to be considered in durability performance of cracked concrete. A crack width/cover depth ratio of 0.01 produced a marginal increase in the D_a in the tension zone of a loaded prism. This relationship also had an effect on the critical chloride content by a hyperbolic relationship.

2.2 Critical chloride content

Service life models for steel reinforced concrete rely strongly on the concept of critical chloride content, that is, the chloride level at the steel surface that causes initiation of corrosion. Substantial research has been carried out over the past years with regard to C_{crit} in uncracked concrete [19-24]. However, high uncertainties are associated with values for the critical content due to poor understanding of the influences of pH and chemical composition of pore liquid, moisture and oxygen (related to environment and cover depth) and the quality of the concrete/reinforcement interface such as the presence of cracks (related to mechanical loads, cement type and curing). Better understanding of important influences can reduce uncertainties; however, corrosion initiation is an inherently stochastic process [25]. In previous work [26], it was shown that results could be used to obtain meaningful statistical distributions. A simplified view is that chloride ions break down the passive film, after which corrosion proceeds without any further influence of chloride [27]. Influx of hydroxyl ions will reduce the activity in the corrosion pit. Dissolved iron ions and a low pH in a pit pull both chloride and hydroxyl ions in, proportionally to the inverse of the concrete electrical resistivity. A minimum concentration of chloride ions in the pore solution is necessary to get the process running. Both availability of chloride and low resistivity are favourable to sustain such an incipient corrosion pit and to develop a high corrosion rate in a short time. At some point, the corrosion rate is so high that repassivation is no longer possible; it becomes independent of chloride content. If the resistivity is high, transport of chlorides may be too slow to sustain corrosion and allow the pit to grow. Presently, there is no widely accepted test method for the critical content. Also, the relationship between crack width and critical chloride content would be valuable for service life modelling.

3 Corrosion propagation in cracked concrete

The propagation of corrosion in cracked concrete is still under debate. Beeby [28] and Arya [10] suggest that there is no direct relationship between crack width and corrosion rate. Supposedly cracks only accelerate initiation but they do not accelerate propagation. However, several factors must be considered: the activity of the crack (i.e. active, that is, varying or dormant), properties of the concrete-steel interface and the environment. Studies have focused on two important factors related to the propagation of corrosion: crack width and crack spacing.

3.1 Crack width

Corrosion at cracks seems to be less severe than expected as long as crack widths are limited to below about 0.5 mm at the concrete surface. Several studies [2-11] have proven that the steel intersected by the crack behaves as anode, while in the uncracked concrete it remains cathodic. Suzuki et al [29] found that the steel inside cracks was depassivated earlier regardless of the crack width. In specimens containing several cracks, the widest became active thus suppressing corrosive action in smaller cracks. Finally, the w/b ratio was found to have greater effect on corrosion than cover depth. In work by Otieno et al. [3], the presence of cracks accelerated chloride-induced corrosion by increasing the penetrability. In blended cements, it was found that the concrete quality (binder type and w/b ratio) could be used to control corrosion propagation. However, a universal threshold crack width was not found since it depends on the properties of each concrete composition. Mangat et al. [30] reported that an important factor controlling steel corrosion in marine environments is permeability of concrete, since it regulates the ingress of Cl^-, O_2 and moisture. The effect of binder type on the corrosion of steel in cracked concrete was studied by Scott et al. [31]. The use of blended cements reduced the corrosion rates of cracked concrete compared to Portland cement concrete under the same conditions.

3.2 Crack spacing

Arya et al. [10] studied the effect of combined crack width and spacing on concrete specimens. Results show that the smaller the frequency of cracks (i.e. the distance between them) the smaller the amount of corrosion. Raupach et al [11], found that the application of a smaller bar diameter in order to limit the crack width resulted in a higher deterioration rate despite the smaller crack distance and crack width. Schießl et al. [6] found that the moisture content was the most influencing factor on the corrosion rate. During wetting, the highest corrosion rates were found. When drying, the corrosion rate was reduced considerably. In another publication, it was found that for each crack width the corrosion rate of reinforcement steel in concrete is rather more influenced by the surrounding concrete [5]. Also, the orientation of the crack was found to be of importance during the propagation phase [32]. Longitudinal cracks were found to be more critical due to the loss of passivity in a larger surface of the reinforcement.

3.3 Macro-cell corrosion

The propagation of corrosion in cracked concrete was found to have macro-cell behaviour. This was achieved with the use of a set-up which contained a segmented bar that performed as anode (inside the crack) and cathode (in uncracked concrete) with a determined area [5, 6, 11]. Mohamed et al. [5], found both micro and macro-cell behaviour in concrete specimens subject to cracks. Schießl et al [6] found that cell current measurements on cracked concrete showed the formation of macrocells.

The steel at a crack behaves as an anode and the steel between cracks, up to a distance of several hundreds of millimetres from the cracks, as cathode. Also, that the corrosion rate is influenced by the conditions between cracks. Increasing the cover depth and reducing the w/b ratio reduced the corrosion rate of steel bars. However, after 2 years no clear relationship between crack width and corrosion rate was seen. Raupach [11] found that the cathode area was extended up to 200 millimetres apart from the crack, and the anodic current was balanced by the total cathodic current. When the distance from the crack increased, the cathodic current on the segmented bar decreased.

3.4 Interfacial transition zone

The interfacial transition zone (ITZ) has been considered as one of the most important parameters when studying steel corrosion in concrete [33]. Disruptions at microscopic level have an important influence on the kinetics of corrosion. However, most of the research carried out has been referred to uncracked concrete where the ITZ has its best properties. The influence of cracks on the disruption of ITZ is interesting due to the subsequent effects that may produce microcracking or local debonding. The thermodynamical conditions of the steel reinforcement may be influenced by this effect as well.

3.5 Corrosion products

The formation of corrosion products on the steel surface is a serious threat for the durability of the structure [34-36]. Corrosion products have a higher volume than iron, producing internal stresses which lead to cracking and spalling. For cracked concrete, Jaffer et al. [36] found that different types of oxides are formed on the steel surface depending on the thermodynamic conditions at the depth of the reinforcement. Ferric hydroxide, which has a volume four times higher than iron, was found in test specimens containing Portland cement. Differences in the composition of corrosion products may be present when using different binders, i.e. slag or fly ash cements. Since most of the civil infrastructure in The Netherlands is constructed with blast furnace slag cement with a high slag percentage (c. 70%, CEM III/B), the influence of cement on corrosion products needs more investigation.

3.6 Mitigation mechanisms

Possible mitigating mechanisms reduce the rate of subsequent corrosion, such as self-healing [37, 38], possibly by stopping chloride ingress and/or causing lack of sufficient chloride to sustain growth of corrosion pits. Another mechanism that has

Fig. 3 Corrosion products and mitigation mechanisms

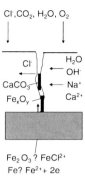

been reported in literature is the diffusion of corrosion products, through the cement matrix [7, 39-41]. These products could also block further ingress of chlorides, increasing the length of the propagation period. This behaviour may be considered when explaining the reduced effect of corrosion in cracked concrete structures exposed to marine environment. Figure 3 shows the possible mitigation mechanisms and formation of corrosion products inside cracks.

4 Conclusions

The performance of steel reinforcement in cracked concrete exposed to chlorides is important for estimating the service life of concrete infrastructure. Until now, structural codes control the allowable crack width with respect to exposure conditions. However, other aspects like concrete composition (binder type), cover depth and the interaction between crack width and spacing should be accounted for. It is necessary then, to understand the mechanisms involved and to study the influence of technological and design parameters on corrosion of steel reinforcement. This could lead to re-evaluating the current codes in order to avoid unnecessary repair. Since corrosion of steel is a stochastic phenomenon, a model for corrosion estimation due to cracks should include the statistical distributions of parameters, i.e. a probabilistic model.

Acknowledgement Financial support by the Dutch Technology Foundation (STW) for project 10978 M3C4 with the framework of Perspectief programma IS2C, as well as its industrial sponsors is gratefully acknowledged.

References

[1] fib, (2006), Model Code for Service Life Design, fib Bulletin 34, Model Code, 116 pages, ISBN 978-2-88394-074-1.
[2] Bouwmeester, W.J. and Schlangen, E., (2008), In: *Tailor made concrete structures*, JC Walraven & D Stoelhorst (Eds.), Taylor & Francis, London, 65–70.

[3] Otieno, M.B., Alexander, M.G. and Beushausen, H.D., (2010), *Mag Concrete Res*, vol. 62, n.6, pp. 393–404.
[4] François R., Arliguie, G., (1998), *J Mater Civil Eng*, vol. 10, n. 1, pp. 14–20.
[5] Mohammed T.U., Otsuki N., Hisada M., (2003), *J Mater Civil Eng*, vol. 13, n.3, pp. 460–469.
[6] Schießl P., Raupach, M., (1997), *ACI Struct J*, vol. 94, n. 1, pp. 56–61.
[7] Bentur A., Diamond S. and Berke N.S., (1997), *Steel corrosion in concrete*, Chapman-Hall.
[8] Mehta P.K., Gerwick B.C., (1982), *Concrete International*, October, pp. 45–51.
[9] Otsuki N., Miyazato S., Diola N.B., Suzuki, H. (2000), *ACI Mater J*, vol. 97, n. 4, pp. 454–464.
[10] Arya C., Ofori-Darko F.K., (1996), *Cement and Concrete Research*, Vol. 26, No. 3, pp. 345–353.
[11] Raupach M., (1996), In: *Corrosion of reinforcement in concrete construction*, C.L. Page, P.B. Bamforth and J.W. Figgs (eds.), Cambridge, 13–23.
[12] Wilkins N.J.M., Lawrence P.F., (1983) In: *Corrosion of reinforcement in concrete construction*, A.P. Crane, (ed.), pp. 119–141.
[13] Bertolini L., Elsener B., Pedeferri P., Polder R.B., (2004), *Corrosion of Steel in Concrete: Prevention, Diagnosis, Repair*, Wiley-VCH Verlag GmbH & Co. KGaA, Weinheim, ISBN 3-527-30800-8.
[14] Win P. P., Watanabe M., Machida A., (2004), *Cement Concrete Res*, vol. 34, p. 1073–1079.
[15] Djerbi A., Bonnet S., Khelidj A., Baroghel-bouny V., (2008), *Cement Concrete Res*, vol. 38, pp. 877–883.
[16] Boulfiza M., Sakai K., Banthia N. and Yoshida H., (2003) *ACI Materials Journal*, V. 100, No. 1, January-February 2003, Title no. 100-M5, pp. 38–48
[17] Konin A., Francois R., Arliguie G., (1998), *Materials and Structures*, Vol. 31, June 1998.
[18] Gowripalan, N., Sirivivnaton, V. and Lim, C.C. (2000), *Cement Concrete Res*, vol. 30, pp. 725–730.
[19] Angst U., Elsener B., Larsen C.K., Vennesland Ø., (2009), *Cement Concrete Res*, 39, 1122–1138.
[20] Tang L., Utgenannt P., (2009), *Materials and Corrosion*, 60, No. 8.
[21] Alonso M.C., Sanchez M., (2009), *Materials and Corrosion*, 60, No. 8.
[22] Alonso, M.C., Andrade C., Castellote M., Castro P., (2000), *Cement Concrete Res*, 30, 1047–1055.
[23] Markeset G., (2009), *Materials and Corrosion*, 60, No. 8.
[24] Nygaard P.V., Geiker M.R., (2005), *Materials and Structures*.
[25] Izquierdo, D., Alonso, M.C., Andrade, C., Castellote, M., (2004), *Electroch. Acta* 49, 2731–2739.
[26] Polder, R.B., Peelen, W.H.A., (2002), *Cement & Concrete Composites*, Vol. 24, 427–435.
[27] Polder, R.B., (2009), *Materials and Corrosion*, 60, (8), 623–630
[28] Beeby A.W., (1983), *Concrete International*, 5, 2, p. 35–68.
[29] Suzuki K., Ohno Y., Praparntanatorn S., Hamura T., (1996), In: *Corrosion of Reinforcement in Concrete*, C.L. Page, P.B. Bamforth and J.W. Figgs (eds.), Cambridge, 19–28.
[30] Mangat P.S., Gurusamy K., (1987), *Cement Concrete Res*, 17, p. 385–396.
[31] Scott A., Alexander M.G., (2007), *Magazine of Concrete Research*, 59, No. 7, p. 495–505.
[32] Darwin D., Manning D.G., Hognestad E., Beeby A.W., Rice P.F. and Ghowrwal A.Q., (1985), *Concrete International*, May, pp. 20–35.
[33] Page C.L., (2009), *Materials and Corrosion*, 60, No. 8
[34] Hansson C.M., Okulaja S.A., (2003), In: *Advances in Cement and Concrete*, University of Illinois at Urbana – Champaign, Copper Mountain
[35] Marcotte, T.D., Hansson, C.M. (2003), *J Mater Sci*, vol. 38, pp. 4765–4776.
[36] Jaffer S.J., Hansson C.M., (2009), *Cement Concrete Res*, 39, 116–125.
[37] Sahmaran, M. (2007), J Mater Sci, vol. 42, pp. 9131–913.
[38] Schlangen E., Joseph C., (2008), In: *Self-healing Materials: Fundamentals, Design Strategies and Applications*, Swapan Kumar Ghosh., (ed.)Wiley-VCH GmbH & Co, Weinheim, 141–182.
[39] Allan M.L., (1995), *Cement Concrete Res*, 35, 6, p. 1179–1190.
[40] Aligizaki K.K., Rooij de M.R., Macdonald D.D., (2000), *Cement Concrete Res*, 30, 12, p. 1941–1945.
[41] Marcotte T.D., Hansson C.D., (1998), In: *International Symposium on High-Performance and Reactive Powder Concretes*

Possibilities for improving corrosion protection of reinforced concrete by modified hydrotalcites – a literature review

Zhengxian Yang, Hartmut Fischer, and Rob Polder

Abstract Modified Hydrotalcites (MHTs) represent a group of technologically promising materials for improving corrosion protection in concrete owing to their low cost, relative simplicity of preparation, and plenty of composition variables. Numerous academic and commercial studies on MHTs have been carried out, but few of them on cementitious materials particularly in exploiting their potential application in corrosion protection of reinforced concrete. In this paper, the corrosion mechanism in reinforced concrete and concrete properties that affect it are briefly inroduced. In addition, the existing knowledge with regard to synthesis and characterisation methods of MHTs, ion exchange within the MHT structure as well as the application of MHTs in cementitious materials were reviewed.

1 Introduction

Due to its versatility and relatively low cost, concrete has become the most widely used construction material in the world. Concrete is a porous and highly heterogeneous composite with features of various dimensions ranging from nanometer-sized pores and calcium-silicate-hydrate (C-S-H) gel to micrometer-sized air voids,

---

Z. Yang (✉)
Materials innovation institute (M2i), Delft, The Netherlands

Materials and Environment, Faculty of Civil Engineering and Geosciences,
Delft University of Technology, Delft, The Netherlands
e-mail: Zhengxian.Yang@tudelft.nl

H. Fischer
TNO Materials Performance, Eindhoven, The Netherlands

R. Polder
Materials and Environment, Faculty of Civil Engineering and Geosciences,
Delft University of Technology, Delft, The Netherlands

TNO Building Engineering & Civil Engineering, Delft, The Netherlands

centimetre size aggregate particles and to steel reinforcement that can be meters in length. Exposed to service enviroment, concrete can be penetrated by corrosive agents (e.g. chemical and microbiological substances), liquid (e.g. water with dissolved ions) or gases (e.g. oxygen and carbon dioxide) through capillary absorption, hydrostatic pressure or diffusion. In addition, freeze-thaw cycles in cold climates may also compromise the protection of reinforcement. All of these factors potentially impose a serious threat on the durability and serviceability of concrete structures [1–3]. Among those factors, the dominant aggressive external influence is the ingress of chloride ions, typically present in de-icing salts and marine environment [4].

Modern service life design approaches aim at providing sufficient concrete cover depth to the reinforcing steel, taking into account its resistance against chloride transport. Traditional standards oversimplify the complexity of the mechanisms involved and provide insufficient performance in aggressive environment. Presently available options for improved corrosion protection are too costly and complicated or insufficiently effective. New generation reinforcement such as stainless steel is much more expensive than ordinary reinforcing (carbon) steel [5, 6]. Cathodic prevention or protection may be effective but both are a special niche expertise and are thus not applied on a wide scale [7]. Coatings on the concrete surface could not last long enough (10-20 years), which causes a cycle of maintenance of its own [5]. Corrosion inhibitors have been proposed but are generally not reliable in terms of long-term efficiency [5, 8].

In the last two decades, more interest has been given to study new compounds able to prevent or stop corrosion and other durability related issues [8–10]. A promising class of materials for use in concrete for improved corrosion resistance is formed by modified hydrotalcites (MHTs). Preliminary work at TNO/AIDICO in the Netherlands has shown that ion exchange occurs between free chloride ions in the pore solution and hydroxyl ions intercalated in MHTs, thereby reducing the free chloride concentration [11], which is equivalent to increased binding of chloride. TNO/AIDICO's work has also shown that certain organic anions with known inhibitive properties may be intercalated, which then can be slowly released, possibly 'automatic' upon arrival of chloride ions. Such inhibition would increase the chloride threshold level for corrosion initiation and/or reduce the subsequent corrosion rate. Less aggressive electrochemical potentials have been observed in simulated pore solution experiments with MHTs as compared to solutions without MHTs [11]. Based on these interesting findings related to MHTs and considering their useful addition to concrete, this communication presents some relevant knowledge which should be considered when attempting to use MHTs as a new additive of concrete for improved corrosion protection. Correspondingly, the current state of research on these subjects is also discussed.

2 Corrosion in Concrete

Steel in concrete is normally in a non-corroding and passive condition. During hydration of cement a highly alkaline pore solution (pH>12.5) facilitates the formation of a passive hydrated oxide/hydroxide film on the surface of the steel. This thin, protective film can effectively insulate the steel from the surrounding electrolytes so

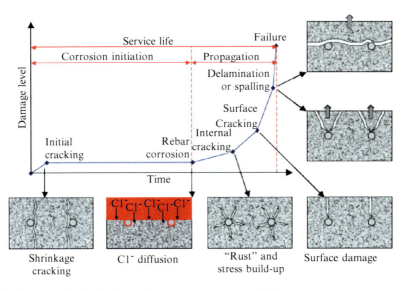

Fig. 1 Corrosion developing in reinforced concrete structure [18]

that the onset of corrosion is delayed [12]. However, this protective film can be disrupted (i.e., depassivation) by the ingress of chlorides and carbon dioxide from the atmosphere (i.e., carbonation). Chloride ion transport by diffusion or capillary absorption in the concrete pore system is relatively fast and deleterious. Furthermore, the gradual ingress of carbon dioxide reacts with solid $Ca(OH)_2$ and other cement hydration products reducing the pH of the concrete pore solution to a value around 9.0 where passivity of reinforcing steel is destroyed. Although corrosion due to carbonation progresses at a lower rate than that due to chloride ingress [13], the combination of chloride ingress and concrete carbonation increases the corrosion risk even more. Simultaneously, in above-ground outdoor structures, plenty of oxygen and water are available to promote relatively rapid corrosion. Besides these two main causes, there are few more factors, some related to concrete quality, such as w/c ratio, cement content, impurities in the concrete ingredients, presence of surface cracks, etc. and others related to the external environment, such as acids, sulfates, freeze-thaw cycles, bacterial attack, stray currents, etc., which may lead to reinforcement corrosion. In the presence of oxygen and moisture, corrosion of steel in concrete is an electro-chemical process which creates a current flow causing metal to dissolve. The reactions involved can be simplified as Eqns. (1) and (2) although they are actually numerous and far more complicated [12–17].

$$\text{Anode reaction: } Fe \rightarrow Fe^{2+} + 2e^- \text{ (and: } Fe^{+2} + 2H_2O \rightarrow Fe(OH)_2 + 2H^+ \text{)} \quad (1)$$

$$\text{Cathode reaction: } O_2 + 2H_2O + 4e^- \rightarrow 4OH^- \quad (2)$$

The development of corrosion in concrete structures can be divided in two main stages according to Tuutti's model as shown in Fig. 1[18, 19]. The first stage is the initiation of corrosion, in which the reinforcement is passive but phenomena that can lead to loss of passivity, e. g. carbonation or chloride attack, take place. The second stage is corrosion propagation, depending on the availability of water and oxygen in the proximity of the steel. It starts when the steel is depassivated and may cause failure of serviceability associated with cracking and spalling of the concrete cover. The time until structure failure is often referred to as the service life, which is determined by the sum of these two stages.

3 Concrete Properties that affect Corrosion of Reinforcement

3.1 *Permeability*

Concrete, as a porous and highly heterogeneous composite, is subject to the ingress of various ionic and molecular species from the environment. The term permeability indicates, in general, the property to allow substances to intrude the concrete and attack the reinforcing steel resulting in corrosion. The penetration of gases, liquids or ions into concrete takes place through pore spaces in the hardened cement paste matrix and paste-aggregate interfaces or microcracks according to four basic mechanisms[4]:capillary suction, permeation (due to pressure gradients), diffusion (due to concentration gradients) and migration (due to electrical potential gradients) [20]. Permeability is believed to be a decisive characteristic of concrete durability, which is related to its microstructural properties, such as the size, amount, distribution, tortuosity and connectivity of pores and microcracks [21, 22].

3.2 *Chloride Binding*

Chloride binding in concrete may be defined as the interaction between the porous matrix and chloride ions which results in effective removal of chlorides from the pore solution [22]. Effective binding will remove a part of chloride from the transport process as well as alter the pore solution concentration and therefore the concentration gradient driving chloride diffusion. The critical chloride content level able to initiate corrosion at the depth of embedded reinforcing steel will be increased correspondingly for concrete with a high chloride binding capacity. Depending on the reaction enthalpy and mechanism, the binding of chloride ion can take place through both chemical combination and physical adsorption. There are many factors associated with the constituents of the concrete affecting the chloride binding capacity. These factors include cement type, curing temperature and age, pH of pore solution, w/c ratio, and chloride concentration and so on [23].

4 MHTs and Their Application in Cementitious Materials

4.1 General Aspect

Hydrotalcite was discovered in around 1842 in Sweden but the first exact formula, [Mg$_6$Al$_2$(OH)$_{16}$]CO$_3$·4H$_2$O was not published until 1915 by Manasse[24]. Subsequently hydrotalcite gave its name to the large mineral group of naturally occurring Layered Double Hydroxides (LDHs) which are also known as hydrotalcite-like materials or commonly referred to as hydrotalcites. Nowadays, numerous pure LDHs compounds have been prepared in the laboratory. Since all of these synthetic LDHs have the same structure as their parent material of hydrotalcite, they can also be termed as modified hydrotalcites (MHTs), a term used throughout this text. MHTs are structurally similar to the minerals Brucite (Mg(OH)$_2$,) and Portlandite(Ca(OH)$_2$), in which a central divalent metal cation is surrounded by six hydroxyl groups in an octahedral configuration. These octahedral units form infinite, charge-neutral layers by edge-sharing, with the hydroxyl ions sitting perpendicular to the plane of the layers. The layers then stack on top of one another to form a three-dimensional structure. In MHTs, a fraction of the divalent cations is isomorphously substituted by trivalent cations, which generates a net positive charge on the layers that necessitates the incorporation of charge-balancing anions in the interlayer galleries. The remaining free space of the interlayer is occupied by water molecules via hydrogen bonding. Although the most common anion found in naturally occurring hydrotalcites is carbonate, in practice, however, there is no significant restriction to the nature of the interlayer charge-balancing anion. The MHT structure represented by a general formula [M$^{II}_{1-x}$ M$^{III}_x$ (OH)$_2$]$^{x+}$ [(A$^{n-}_{x/n}$)]$^{x-}$·mH$_2$O can accomodate various cations in the hydroxide layers with varing M^{II}/M^{III} ratios as well as a great variety of anionic species in the interlayer regions. The x value is in the range 0.22-0.33. A typical structure of MHTs can be schematically shown as in Fig. 2.

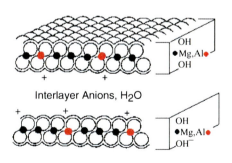

Fig. 2 Schematic diagram of a typical MHTs structure

In cement chemistry, MHTs represent one of the phases that form during cement hydration. The hydration products of tricalcium aluminate (C_3A) and tetracalcium aluminoferrite (C_4AF) are hexagonal-layered materials denoted by $C_2(A,F)H_8$, $C_4(A,F)H_{13}$, and $C_4(A,F)H_{19}$. These hydrates, along with the AFm phases (tetracalcium aluminate monosulfate) all can be seen as hydrotalcite-like compounds [10]. It is claimed that the stability of AFm phases plays a very important role in controlling the performance of concrete since chloride ions may interact with hydrated cement forming chloroaluminate phases, such as Friedl's (a chloride-bearing AFm phase) and/or Kuzel's (a chloride- and sulfate-bearing AFm phase) salt and solid solutions with other AFm phases [25,26]. Formation of these phases, containing essentially chloride, is potentially a mechanism for retarding chloride diffusion and thus mitigating chloride-induced corrosion.

4.2 Synthesis and characterization

A wide variety of cation combinations (e.g., M^{II}, M^{III}) and different types of anions in the interlayer are found in synthetic MHTs compounds. It is suggested that only M^{II} and M^{III} ions having an ionic radius not too different from that of Mg^{2+} may be accommodated in the octahedral sites of the close-packed hydroxyl ions in brucite-like layers to form LDH compounds [27]. The most reliable composition range corresponds approximately to $0.2 \leq x \leq 0.33$ [28]. A number of synthetic methods has been successfully conducted to prepare MHTs; however, three main methods are frequently used. The most commonly used are simple coprecipitation of two metal salts in alkaline solution at constant pH values of about 10. The second one is based on the ion exchange process in which guest ions are exchanged with the anions present in the interlayer spaces of preformed LDHs to produce specific MHTs containing the target anions. The third method is reconstruction which is based on the so-called "structural memory effect". A common concern with all the methods is that in preparations of MHTs with anions other than carbonate it is important to avoid contamination by CO_2 since the carbonate anion is readily intercalated and tenaciously held in the interlayer. More details about the synthetic techniques for the preparation of MHTs can be found elsewhere [28, 29].

Various modern techniques have been employed to characterise MHTs. X-ray diffraction (XRD) and infrared spectroscopy (FTIR) are principally performed to check the structural characteristics of MHTs. Furthermore, energy dispersive X-ray spectroscopy (EDX) [10], nuclear magnetic resonance (NMR) [30, 31], atomic absorption spectroscopy [32], ion coupled plasma optical emission spectrometer (ICP-OES) [33, 34], ion chromatography (IC) [33] are also carried out to determine the compositional formulae of MHTs. The thermal behaviour of MHTs is commonly studied using thermogravimetry (TG), and differential scanning calorimetry (DSC). Morphology and particle size are often checked by electron microscope (e.g., SEM, TEM, FESEM) and particle size analyzer.

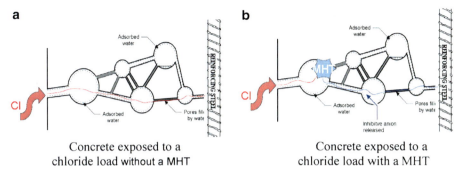

Fig. 3 The role of MHTs in reinforced concrete exposed to chloride load

4.3 Ion Exchange of MHTs and Its Role in Capturing Chloride

A key feature of MHTs is their high anionic exchange capacity ranging from 2 to 4.5 meq/g and therefore the ability to capture a wide range of organic or inorganic anions. The anion-exchange capacity of MHTs is inversely dependent on the layer charge density (i.e., the M^{II}/M^{III} molar ratios). Additionaly, it is affected by the nature of the interlayer anions initialy present. MHTs have greater affinities for multivalent anions relative to monovalent anions [27, 28]. The ion-exchange equilibrium constant tends to increase as the diameter of the anion decreases [35]. Previous research [36] has also found that the interlayer interactions can be directed or mediated through other species present in the interlayer region by coulombic forces between the positively charged layers and the anions in the interlayer and by hydrogen bonding between the hydroxyl groups of the layer with the anions and the water molecules in the interlayer. In summary, the MHTs have a rich interlayer chemistry and participate in anion exchange reactions with great facility. For the envisaged use as an additive to concrete against chloride attack, certain inorganic or organic anions with known inhibitive properties may be intercalated in the structures of MHT's, which then can be slowly released, possibly 'automatic' upon arrival of chloride ions. The anion exchange process can be described according to Eqn. (3) and also shown schematically in Fig 3:

$$HT\text{-}Inh + Cl\text{-}(aq) \rightarrow HT\text{-}Cl + Inh-(aq) \qquad (3)$$

Where *Inh* represents the intercalated inorganic or organic inhibitive anions.

4.4 Application in cementitious materials

Previous studies [37–45] have demonstrated that several ion exchangers based on hydrotalcite (HT) dispersed in polymeric coatings are strong corrosion inhibitors for several metallic substrates. The corrosion is hindered due to the release of an inhibitor anion that diffuses through the pore space of the coating, and it is exchanged for chloride ions in the environment, which are 'captured' in the HT structure [37, 38]. These results suggest that MHTs have a high potential as active component in concrete with corrosion protection properties that can be tailor-made.

Sustainable development of concrete infrastructure is a major challenge in today's construction industry. The use of supplementary materials or functional additives is an integral component of these strategies. Hydrotalcite or hydrotalcite-like phases have been found in hydrated slag cements, which are known to bind more chloride ions than pure Portland cements [46–49]. Nonaka and Sato [50] patented a cement modifier containing hydrotalcite or hydrotalcite and blast furnace slag and provided a method capable of blending this modifier into to concrete/mortar without inhibiting the hydration reaction of cement and a method for imparting required corrosion resistance and durability into the cement/ mortar. The existence of hydrotalcite-like phases such as Friedl's salt (see above) or its iron analogue has been believed to contribute to chloride binding and thus enhances the corrosion resistance of reinforced concrete. The beneficial effects of Friedl's salt on binding chloride in cement support the idea of using other MHTs in concrete as an effective chloride capturer or scavenger. Increased binding would definitely slow down chloride transport. Furthermore, MHTs are expected to be able to retain bound chloride even in a neutral and high temperature environment, for example if the pH of the pore solution drops due to carbonation [35]. Raki et al. [10] demonstrated the promise of MHTs as suitable hosts for intercalation of organic admixtures with the long-term view of controlling their release rate in concrete by blending the inorganic-organic nanocomposites with the cement. In their study, nitrobenzoic acid (NBA), naphthalene-2,6-disulfonic acid (26NS), and naphthalene-2 sulfonic acid (2NS) salts, were intercalated through anion-exchange of nitrate in the host material, $[Ca_2Al(OH)_6]$ $NO_3 \cdot nH_2O$. It was suggested that potential future applications of these composites could be to control the effect of admixtures on the kinetics of cement hydration by programming their temporal release. In addition, a recent patent [51] discloses that a controlled release formulation for a cement-based composition can be produced by means of hydrotalcites modified with different anions, which confer functions such as an accelerator, a set retarder, and a superplasticizer when released. The patent also mentions the potential corrosion inhibition function of this formulation but no examples are supplied. Japanese patents [52,53] disclose a cement admixture that comprises hydrotalcite and other functional additives and it is claimed that the hydrotalcite-based admixture has excellent ability to capture chloride or carbonate ions, thus can simultaneously prevent concrete deterioration caused by salt and carbonation over a wide range of fields. Tatematsu et al. [9] systhesized a hydrocalumite-like material (a calcium-based MHT) and added it to cement mixtures as a salt adsorbent. Their results showed that once the admixed salt adsorbent contacts with chloride ions it could adsorb them and release the intercalated nitrite anions in the meantime. The released nitrite ions on the other hand could work as corrosion inhibitor. The corrosion inhibiting effect of the salt adsorbent on chloride-induced corrosion was further confirmed by experiments performed with a large-size specimen. In a patent, Tatematsu et al. [54] also disclosed a hydrocalumite that contains nitrite or nitrate ions which can be added directly to the concrete as a chloride ion scavenger to prevent choride-induced corrosion. Kang et al. [55] patented a corrosion inhibitive repair method of reinforced concrete structure using repair mortar containing nitrite-based hydrocalumite and confirmed its effectiveness in inhibiting corrosion. A cement additive for inhibiting concrete deterioration was developed with a mixture of an inorganic cationic exchanger: a calcium-substituted zeolite capable of absorbing alkali ions (e.g., sodium, potassium) and an inorganic anionic exchanger: hydrocalumite capable of exchanging anions (chlorides, nitrates, sulfates, etc.) that is directly incorporated in mortar or concrete [56, 57]. The results of their tests indicated the potential of protecting concrete structures from deterioration by exchange of alkali and chloride ions to mitigate alkali-aggregate reaction and corrosion of reinforcing steel. It may be noted that corrosion inhibition systems based on this cement additive refer to hydrotalcites intercalated with inorganic anions, but the possiblity of using organic anions as the interlayer material is not discussed. Another Japanese patent [58] describes on one hand, the use of a carbonate hydrotalcite as a system able to incorporate chloride ions from the environment in the HT structure, and on the other hand, polycarboxylic acid and organoamine as corrosion inhibitors which are also added to the cement-based materials to prevent carbonation and corrosion of reinforcing steel bar due to the penetration of water, oxygen, CO_2 etc., as well as to adsorb and remove salt contained in the concrete. However, this work does not deal with the use of organic anions intercalated in the HT structure.

5 Concluding Remarks

Corrosion-related durability problems cause high maintenance costs and safety issues, so any slight improvements in the design, production, construction, maintenance, and materials performance of concrete could be crucial in improving the service life of concrete structures and thus may have enormous social, economic and environmental impacts. MHTs represent one of the most technologically promising materials owing to their low cost, relative simplicity of preparation, and plenty of composition/preparation variables that may be adopted. Up to date, a great deal of work based on academic and commercial interest on MHTs has been carried out, but few of them concern cementitious materials particularly in exploiting their potential application in corrosion protection of reinforced concrete structure. This has led us to initiate additional research. We are confident that future work on applications of new concrete additives based on MHTs will expand rapidly and contribute greatly to the effort of searching for effective measures to improve the durability of reinforced concrete.

Acknowledgements The research was carried out under the project number M81.609337 in the framework of the Research Program of the Materials innovation institute (M2i) (www.m2i.nl). The authors would like to extend their thanks to Mr. J.M. Lloris Cormano and José Caballero for their previous effort put on this project.

References

[1] Arockiasamy, M. (2000), *Sum of Final Report SRT#0510847*, Dept. of Trans. FL.
[2] Yunovich, M. and Thompson, N.G. (2003), *Concr. Int.*, Vol. 25, n.1, p. 52.
[3] Küter, A. (2009), *Ph.D. thesis*, Technichal University of Denmark, Lyngby.
[4] Bertolini, L., Elsener, B., Pedeferri, P. and Polder, R.B. (2004), *Corrosion of Steel in Concrete: Prevention, Diagnosis, Repair*, Wiley-VCH, Weinheim.
[5] Cigna R., Andrade C., Nürnberger U., Polder R.,Weydert R. and E. Seitz (Eds.) (2002), *COST 521: Final Report*, Luxembourg.
[6] Elsener, B., Addari, D., Coray, S., and Rossi, A. (2010), *Mater. Corros.*, vol.61, p.1.
[7] Pedeferri, P. (1996), *Constr. Bldg. Mater.*, Vol.10, n.5, p.91.
[8] Elsener, B. (2001), *Corrosion Inhibitors for Steel in Concrete-State of the Art Report*, EFC Publication No. 35, IOM Communications, London.
[9] Tatematsu H. and Sasaki T. (2003), *Cem. Conc. Res.*, Vol.25, n.1, p.123.
[10] Raki L., Beaudoin J.J. and Mitchell L. (2004), *Cem. Conc. Res.*, Vol.34, n.9, p.1717.
[11] Lloris Cormano, J.M. (2006), *internal TNO Report*, Delft, The Netherlands.
[12] Gaidis J.M.(2004), *Cem. Conc. Res.*, Vol.26, n.3, p.31.
[13] Basheer, L., Kropp, J. and Cleland, D. J. (2001), *Constr. Bldg. Mater.* Vol.15 n.2, p.93.
[14] Mehta P. K. and Monteiro P.J. M. (1993), *Concrete: Structure, Properties and Materials*, Prentice-Hall Inc., Englewood Cliffs, New Jersey.
[15] Rosenberg A., Hansson C.M. and Andrade C. (1989), In: *Materials science of concrete I*, p. 285-313, Skalny JP (Ed), The American Ceramic Society, US.
[16] Kruger J. and Calvert J.P. (1967), *J. Electrochem. Soc*, vol.144, n.1, p.43.
[17] Moreno M., Morris W, and Duffo G.S.(2004), *Corros Sci*, vol46, n.11, p.2681.
[18] Daigle L., Cusson D. and Lounis, Z. (2008), In: *Shrinkage and Durability of Concrete and Concrete Structures (CONCREEP 8)*, pp.1-6, 8th Intl Conf. on Creep, Japan.
[19] Tuutti K.(1982), Swedish Foundation for Concrete Research (CBI), Stockholm.
[20] Stanish, K.D., Hooton, R.D. and Thomas, M.D.A. (1997), *FHWA contract DTFH61-97-R-00022 report*, University of Toronto, Canada.
[21] Savas, B.Z. (1999), *PhD thesis*, North Carolina State University, Raleigh, NC.
[22] Glass, G.K. and Buenfeld, N.R. (2000), *Prog. Struct. Engng Mater*, vol.2, n.4, p.448.
[23] Glass, G.K. and Buenfeld, N. R.(1997), Mag Concr Res, vol.49, n.181, p.323.
[24] Manasse, E. (1915), *Atti Soc. Toscana Sci. Nat.*,vol.24, p.92
[25] Balonis M. (2010), *PhD Thesis*, University of Aberdeen, UK.
[26] Csizmadia, J., Balazs, G. and Tamas, F.D. (2001), *Cem. Conc. Res.*, vol.31, n.4, p.577.
[27] De Roy, A., Forano, C., Besse J.P. (2001), In: *Layered Double Hydroxides: Present and Future*, pp. 8, Rives, V. (Ed). Nova Science Publishers Inc, New York.
[28] He, J., Wei, M., Li, B., Evans D.G. and Duan X. (2006), *Struct Bond*, vol.119, p.89.
[29] Kanezaki, E. (2004), *Interface Science and Technology*, vol.1, p.345.
[30] Carlino, S. (1997), *Solid State Ionic*, vol.98, n.1, p.73.
[31] Rey,F. and Forne´sV.(1992), J. Chem. Soc., Faraday Trans., vol.88, n.15, p.233.
[32] Whilton, N.T., Vickers, P.J. and Mann, S.J. (1997), *J. Mater. Chem.*,vol.7, n.8, p.1623.
[33] Rozov, K., Berner U., Taviot-Gueho, C. Leroux, F., Renaudin, G., Kulik, D. and Diamond, L.W. (2010), *Cem. Conc. Res.*, vol.40, n.8, p.1248.
[34] Kang, M.R., Lim H.M., Lee, S.C., Lee, S-H. and Kim, K.J. (2004), *Adv Technol Maters Process J*, vol.6, n.2, p.218.
[35] Miyata, S. (1983), *Clays Clay Miner.*, vol.31, n.4, p.305.
[36] Taylor H.F.W.(1973), *Miner Mag.*, vol.39, p.337.
[37] Buchheit, R.G., Guan, H. and Wong, F. (2003), Prog. Org. Coat. vol.47. n. 3-4, p.174
[38] Williams, G. and McMurray, H.N. (2003), Electrochem Solid St. vol.6, n.3, p.B9.
[39] Worsley, D.A. (2003), *World Patent*, NO. WO03/102085.
[40] Miyata, S. (1988), *US Patent*, NO. 4761188.
[41] Miyazawa, S., Morihira, Y. and Fujimori, H. (2002), *US Patent* NO. 6383270.

[42] Gichuhi, T. and Novelli, W. (2007), *EP Patent*, No. EP1591493A3.
[43] Gichuhi, T. and Novelli, W. (2005), *US Patent*, No. US2005235873A1.
[44] Sinko, J., and Kendig, M.W. (2005), *US Patent*, No. US2005022693A1.
[45] Abe Y, Chiba, M, and Uchiyama, H. (2005), *Japan Patent*, No. 2005336002.
[46] Arya, C., Buenfeld, N.R., Newman, J.B. (1990), *Cem. Conc. Res.*, vol.20, n.2, p.291.
[47] Dhir, R.K. and Dyer, T.D. (1996), *Cem. Conc. Res.*, vol.26, n.12, p.1767.
[48] Arya, C. and Xu, Y. (1995), *Cem. Conc. Res.* vol.25, n.4, p.893.
[49] Glass, G. K. and Buenfeld, N. R. (2000), *Corros. Sci.*, vol.42, n.2, p.329.
[50] Nonaka, S. and Sato, T. (2001), *Japan Patent*, No. JP2001089211A2.
[51] Raki, L., and Beaudoin, J.J. (2007), *US Patent*, No. US20070022916A1.
[52] Mihara, T. and Morioka, M. (1993), *Japan Patent*, No. JP05330876A2.
[53] Ashida, K., Handa, M. and Morioka, M. (1993), *Japan Patent*, No. JP05262546A2.
[54] Tatematsu, H., Nakamura, T., Koshimizu, H. and Takatsu, S. (1992), *Japan Patent*, No. JP04154648A2.
[55] Kang, S.P., Kim, G.D. and Hong, S.Y. (2005), *Korea Patent*, NO. KR100515948B1.
[56] Tatematsu, H., Nakamura, T., Koshimuzu, H., Morishita, T. and Kotaki, H. (1995), *US Patent*, No. US 5435846A.
[57] Feng, N. (2007), *China Patent*, No. CN1948207A.
[58] Kashima, M. (1997), *Japan Patent*, No. JP09142903A2.

Determination of the probable failure mechanisms and service life of offshore concrete gravity structures in the OSPAR Maritime Area - research proposal

Rod Jones, Moray Newlands, and Chris Thistlethwaite

Since 1973 concrete gravity structures have been used to extract oil and gas within the OSPAR Maritime Area. As oil and gas supplies are depleting these structures will require decommissioning over the next 10 to 20 years. During design and construction, removal was rarely considered and it is anticipated that this can be extremely high in cost and safety risk. Leaving the structure in situ is a consideration of which there are concerns about the service life of the concrete. Little is known about the durability and failure mechanisms of such structures. Previous research to determine the most probable service life and failure mechanisms is limited as the majority has focused on atmospheric and splash zones, considered to be worst case.

Structures completely submerged are assumed to have a much longer propagation period than structures in splash or tidal zones. This may indicate that established ageing factors such and long-term extrapolation using current model codes need to be evaluated and may not be applicable.

The proposal includes two main phases of work: experimental modelling and probabilistic modelling. Experimentally; oxygen diffusion and corrosion current in un-cracked, statically cracked and dynamically cracked reinforced concrete are to be tested in conditions of artificially created seawater with low oxygen concentration to simulate fully submerged concrete. CEM I concrete is to be tested along with admixtures including fly ash, silica fume and slag admixtures as well as varying w/c ratios and a mixed range of aggregates. Temperature, oxygen concentration and seawater composition will remain constant, replicating a North Sea environment.

With improved data on oxygen availability and corrosion cells in concrete in deoxygenated conditions a probabilistic model of failure can be developed taking into account localised and general corrosion, cyclic and abnormal loading and environmental changes can be developed. This research aims to validate

R. Jones • M. Newlands • C. Thistlethwaite (✉)
Concrete Technology Unit, University of Dundee
e-mail: c.j.thistlethwaite@dundee.ac.uk

and improve previous models and bring together an integrated model for fully submerged concrete, which could benefit a large number of offshore and coastal structures.

1 Introduction

Production of oil and gas in the North Sea has been active since the early 1970s. There are around 600 platforms in this area consisting of four types: small steel platforms (under 4000 tonnes), large steel platforms, concrete gravity based structures and floating platforms [1]. The OSPAR Maritime Area covers the UK Continental Shelf (UKCS), the Norwegian Continental Shelf (NCS) and the Dutch Continental Shelf (DCS). Early estimates of the decommissioning costs vary from £10 billion to £25 billion for the UKCS.

Of the structures, concrete gravity base structures (GBS) have been used extensively in oil and gas production around the world. There are a large number of these offshore structures, of which twenty-seven are located in the OSPAR Maritime Area, with twelve on the UKCS. As the production of oil and gas is beginning to cease decommissioning dates over the following thirty to forty years are planned [2]. The futures of these structures are currently unknown and decommissioning dates continue to be altered as technology to remove oil and gas becomes more readily available. Thirteen of these structures were not initially designed for removal and therefore it is being considered whether leaving the structure in situ is a viable option – environmentally and commercially. This decommissioning of offshore platforms in the North Sea must comply with OSPAR regulations [3] and government guidelines [4].

Four alternative methods are to be considered for decommissioning offshore concrete platforms: float and recycle onshore, float and sink in deep ocean, cut legs to -55m LAT or leave in situ with a navigation aid. All four options have advantages and disadvantages and the option with the lowest cost in terms of safety and financially will be the preferred option. To be able to consider the option of leaving the platform in situ, either cut to -55m LAT or in situ the degradation processes and overall failure of the platform needs to be better understood [5].

Long term durability of concrete in a marine environment is extremely difficult to determine, with many variables affecting the service life of a reinforced concrete structure. To be able to leave the structure wholly offshore, the possibly preferred option due to the cost and safety issues involved in removal, the structure should be shown to not be a hazard to seafaring vessels and the environment. Approval must be acquired from the governing authorities on any decommissioning proposals providing this research opportunity into the long term durability of structures left in situ.

At present, there is little experience on the long term durability of offshore concrete structures in the North Sea, with the first platform constructed in 1973 at the Ekofisk field, and therefore only around 40 years of deterioration has been monitored. Information on the quality of construction and material properties can be difficult to obtain as documents have been lost and misplaced over the years. The durability and failure mechanisms of such structures are as yet unknown and research to determine the most probable service life and failure mechanisms are limited. Research is required to further understand how and when such structures will fail. This research will focus on submerged concrete in a low oxygen environment; replicating the environment that the majority of the shafts and caissons encounter.

2 Background

The service life estimation of a concrete structure will be a function involving many variables, including chloride diffusion, oxygen diffusion, carbonation, loading, corrosion mechanism, etc. This can become very complex and be unrealistic to use within industry. To determine the probability of failure of a structure or the expected age of failure consideration must be made for the loading applied and the structural capacity.

Loading on offshore structures is a complex combination of working, tidal, wave, thermal and wind. Load cases are to be considered in a further study at the University of Leeds, where macro effects on the structure are being considered. Structural capacity is partially affected by the materials used, in this case pre-stressed reinforced concrete forming the gravity base structure with steel topsides. During decommissioning the steel topsides will be removed leaving just the concrete base, therefore consideration will be given to the deterioration of concrete ignoring macro loading.

Currently determination of expected service life is generally based on models of the chloride ingress and duration to initiation. A number of institutions have attempted to quantify this service life deterministically, by quantifying the diffusion coefficient, ageing factor and surface chlorides concentration, a number of which are mentioned in *Table 6.14, pp. 180* [6]. This approach is very simplistic and does not apply any variance to the material properties and environmental aspects. To apply a service life prediction to a structure in industry the capacity of the structure needs to be predicted. Simplistically this will include understanding the corrosion of the steel, which will rely on resistivity of concrete, and oxygen and chloride diffusion into the concrete.

2.1 Corrosion

It is generally agreed that corrosion can be modelled in two phases; initiation and propagation [7-9].

Initiation time, t_i, is the duration of which it takes for corrosion to occur, often due to the chlorides penetrating through the concrete cover and reaching the reinforcement. When the concentration of chlorides at the steel surface reaches the corrosion threshold, i.e. the concentration at which corrosion starts, then the corrosion mechanism moves into a propagation phase. Chloride penetration is likely to be the most important factor in corrosion initiation in a marine environment and is reviewed later.

During the propagation phase electrochemical reactions occur at an anode and cathode. At the anode corrosive products form on the surface of the steel as the metallic ions most commonly Fe^{2+} are released along with two electrons [10]. The cathodic reaction is most commonly the breakdown of oxygen and water using a supply of electrons into hydroxide ions (OH^-). The hydroxide then reacts with the metallic ions producing corrosive products [9].

Some work on the influence of oxygen on corrosion of steel appears to indicate that the diffusion of oxygen is a limiting factor on the corrosion of concrete structures in a fully saturated state [11]. It is thought that once all remaining oxygen within the concrete is consumed the corrosion rate will be only influenced by the oxygen concentration of the surroundings and the diffusion coefficient. Although this is believed here, there is little work to support this scenario. It is therefore required to investigate corrosion in low oxygen environments, as the majority of research appears to concentrate on splash and tidal zones, rather than the submerged area of concrete specific to an offshore platform.

2.2 Oxygen Diffusion

The oxygen at the cathode in a corrosion cell can be the limiting factor in saturated concrete. It is therefore of importance to understand the transport of oxygen through the concrete cover. The most significant research on the transport of dissolved oxygen through concrete was completed in 1986 [12], however the testing was limited to a small number of concrete mixes using only CEM I. Since then further studies have been completed by [13-16]. It was concluded that the range of diffusion for oxygen transport through concrete is between 10^{-5} and 10^{-6} cm^2/s. In contrast [13] report values in the region of 10^{-8} cm^2/s, however it was noted that reported values have ranged from 10^{-6} to 10^{-9} cm^2/s. Such a large variation of diffusion rates is likely to have a large effect on the rate of general corrosion, especially in areas of low concentration of oxygen.

Thickness has been shown to have a negligible effect on the flux of oxygen through concrete, with reduction factors of 2.6 [12] and nothing [15]. However, when predicting the effects of a variable cover over a long period of time, a reduction factor of around 2.6 could affect the overall corrosion rate significantly depending on the importance of the oxygen diffusion factor. A water cement (w/c) ratio of 0.6 is still significantly higher than concrete with a w/c ratio of 0.4, indicating that the aggregate bonding within concrete has a large effect on the flux of oxygen. The ratio between oxygen diffusion and chloride diffusion increases as the w/c ratio decreases as noted by [14].

It has been suggested that in an ocean environment it is likely that the oxygen diffusion coefficients will be reduced further by marine growth and ionic interactions in the surface layer that can cause a reduction in porosity [12]. This surface layer of cement paste was shown, by comparing concrete with two cast faces and two cut faces, to significantly affect the diffusion of oxygen through concrete.

There is an apparent lack of research on oxygen diffusion through submerged concrete, with a small number of differing concrete mix designs being tested. There is a requirement for further work on this area to be able to determine the oxygen diffusion in concrete in cracked and un-cracked sections as well as through a larger range of concrete types, including the effects of varying the aggregate size and type. Any future experimental work should also look to quantify a variance on the results, allowing a probabilistic estimation on oxygen concentrations and diffusion coefficients to assist in modelling the corrosion of steel.

2.3 Chloride Ingress

Chloride ingress is the most prominent cause of deterioration of steel in concrete in the marine environment and has been studied and researched extensively [17-20].

Generally the basis for diffusion of chlorides into concrete is expressed as a variation of Crank's solution to Fick's Second Law and shown in Eqn. 2 [18].

$$C(x,t) = C_s \left(1 - erf\left(\frac{x}{2\sqrt{Dt}} \right) \right) \quad (1)$$

However, diffusion coefficients and surface chlorides concentrations are unlikely to be constant, and the diffusion coefficient is commonly given the form:

$$D(t) = D_{ref} \left(\frac{t_{ref}}{t} \right)^m \quad (2)$$

There are a number of test methods available to determine the ingress of chlorides into concrete. From these test methods, numerous modelling techniques have been used in an attempt to determine the rate of penetration of chlorides [18,21-24].

A fully submerged structure in a stable environment will have a more consistent chloride surface concentration than bridges. Terrestrial bridges can be damaged by air born salts from sea spray, not necessarily on the coast, and de-icing salts commonly used on the roads when the temperature drops to just above degrees Celsius and below. In considering long term chloride surface concentrations for such structures the climate plays an important role, i.e. during a severe winter a substantially larger volume of de-icing salts will be applied to the road network.

Combined research undertaken on chloride ingress in North Sea platforms in the Norwegian sector was completed in 2010 [25]. All data but that from one platform shows that the diffusion coefficients are too scattered to determine a realistic ageing factor for concrete in this marine environment. Data from the structures however was from a range of atmospheric, tidal and submerged zones on varying places on the structures. It was concluded that the most important factors affecting the service life of concrete structures is the localised weaknesses such as cover not being compacted well or not as specified, poor grouting of pre-stressing tendons and other errors made during construction; rather than the properties of the concrete itself.

Ageing Effect on Diffusion. Life-365 [26] gives an ageing factor using the following Eqn. 3 and is commonly used throughout other research:

$$m = 0.2 + 0.4(^{\%FA}/_{50} + ^{\%GGBS}/_{70}) \tag{3}$$

Tests on three samples [27]; Portland cement, 26% fly ash and 70% slag provide ageing factors calculated using best fitting coefficients. These were 0 and 0.12 for Portland cement, 0.8 and 0.58 for fly ash, 1.0 and 0.99 for slag. Comparing these determined values with Life-365; 0.2, 0.41 and 0.6 for PC, FA and Slag respectively it is shown that they can commonly be up to 100% off the measured value.

Other investigations use GGBS [28]. Using *Life-365* the ageing factor is 0.31, *ChloDif* provides 0.1, and DuraCrete produces 0.71 (submerged), 0.6 (tidal/splash) and 0.85 (atmospheric). The experimental results average an ageing factor of 0.47 through a fitted curve. Results show after 13 years, D_{app} has a spread between 2×10^{-11} and 2×10^{-13} m^2/s, a factor of 100.

Data on the *Heidrun* platform [25] shows an ageing factor of 0.82; however this data is only taken from a few cores after 2, 5 and 9 years. *Heidrun* was also constructed to a design strength of LC-60 and contains lightweight aggregate and silica fume, which is expected to have a higher ageing factor than CEM I such as ageing factors of 0.7 for fly ash and 1.2 for slag [18], and it is assumed that silica fume is likely to be similar. However for CEM I it is reported to be only 0.1, which is a comparatively small effect on the diffusion coefficient. Other research [29,-31] has shown ageing factors ranging from 0.2 to 0.6 for CEM I materials, with the majority between 0.2 and 0.35.

Although these values, mainly from samples in laboratories or test specimens in the field, are available and produce a wide scatter of results, there is limited data on marine concrete over a period of time long enough to realistically predict the long term behaviour of concrete under this environment. As chloride ingress is such a vital factor affecting corrosion and ultimately the service life of structures, a greater understanding of the ageing of concrete is required. As it is understood, work is currently being undertaken in an attempt to improve this understanding.

As the ageing factors for CEM I are between 0 and 0.12, with predicted factors of 0.2 and higher, it appears that models in general could be overestimating the ageing effect of CEM I concretes. The main limitations of the research discussed are the short duration of any experimental works along with the method by which the ageing factor is calculated. This factor depends on measurements of chloride diffusion coefficients and surface concentrations, previously shown to have large coefficients of variation.

Effect of Cracking. Cracking occurs in the majority of concrete structures and in design codes is limited for structural, durability and aesthetic reasons. The effects of the cracking on long term durability have been researched extensively as will be discussed. As cracking occurs in most structures under stresses, experimental work must take cracking into consideration to ensure it is applicable to the 'real world' situations.

D_{cr}, the diffusion coefficient through the crack, increased up to a crack width of around 80μm, and other research [32, 33] shows that for an increased crack width that is larger than 75μm, D_{cr} becomes constant. Similarly results conclude the value to be around 80μm [34].

It is also shown that with increases in crack width the chloride diffusion coefficient through the entire specimen, D_e, increases for ordinary concrete and high strength concrete [32]. After cracks become larger than 80μm, the increase in D_e becomes more rapid.

Diffusion of chlorides in concrete is not just increased at the site of the crack but there is an increase in the diffusion coefficient adjacent to the cracking.

Offshore structures are designed for minimal cracking, if the concrete in the tensile face of the leg allows for faster diffusion due to the breakdown of the aggregate/cement interface or micro-cracking the increase in diffusion coefficient must be taken into account in service life estimation.

3 Experimental Procedure

3.1 Oxygen diffusion

Oxygen diffusion testing is to be carried out in two cell testing setup in *figure 1*. This is a variation of the method carried out in two different methods outlined by [12]. Either of these different arrangements may be used, depending on the availability of equipment and available space. Only one method outlined is appropriate to measure the bulk transport properties through concrete as one requires a cast fact against a metal plate.

It is proposed that two scenarios are monitored; cut/cast faces and cut/cut faces to validate the effect of the denser layer adjacent to the cast face. Concrete cylinders are to be cast with a diameter of 100mm and cut into the sections required of varying thickness.

PVC tubes are to be used to encase the concrete and are to be constructed to 150mm length and a suitable thickness, noted as 3mm however subject to change. Concrete specimens of diameter 100mm and variable thickness are then sealed into place with epoxy to prevent any leaking of water, calcium hydroxide around the concrete.

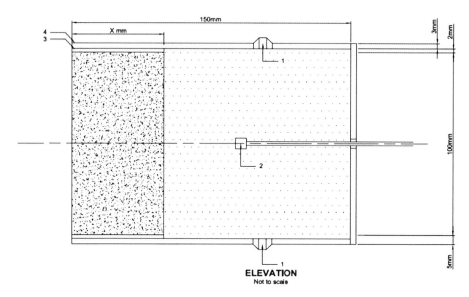

Fig. 1 Elevation of experimental setup

Calcium hydroxide is to be used instead of water within the casing to act as an electrolyte to complete the oxygen sensor. Two electrodes, likely to be stainless steel and graphite, are to be connected to a potentiostat and used to monitor the oxygen concentration within the solution.

To deoxygenate the Calcium Hydroxide, Nitrogen will be bubbled through to displace the Oxygen and lower the concentration of oxygen to negligible values. Each batch of concrete is to be trialled simultaneously - submerged in a large water bath with a controlled temperature, oxygen concentration and composition of seawater to replicate the in situ conditions of the North Sea.

3.2 Corrosion under low oxygen environment

The proposed test to evaluate the corrosion of reinforcement in the concrete specimens uses an adapted methodology from the setup used previously to determine the chloride threshold of reinforcing steel [35]. The following proposal uses a similar method to drive chlorides through the specimen to initiate corrosion, however in this study the propagation of corrosion is to be monitored.

This methodology allows for acceleration of the chlorides reaching the embedded steel allowing for the critical chloride content to be reached causing depassivation of the steel. Once this has occurred the concrete will be subjected to conditions replicating the North Sea and the reinforcement will corrode naturally in an environment with accessible chlorides and lack of oxygen. The corrosion current will be monitored using a linear polarization method to determine the rate of corrosion of the steel. Temperature and composition of the artificial sea water will be controlled.

Fig. 2 Cross section of experimental setup

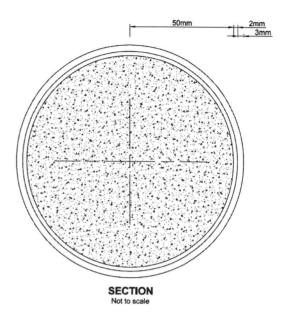

It is proposed that this experiment is run in three phases; un-cracked, statically cracked and dynamically cracked to determine the rate of corrosion in the conditions of concrete experienced in situ. The specimens to be used are 300mm long, 70mm wide and 125mm deep. Two steel reinforcing bars will be placed within the concrete, allowing 25mm cover to the top and bottom of the beam. Tolerance of the cover concrete is not vital during this experiment as the diffusion of chlorides to the surface will be forced, and the propagation is to be monitored outlined as a method.

To initiate corrosion, chlorides will be forced through the concrete using a ponding method. 1M NaCl solution will be placed on the top surface of the specimen housed in a Polyvinyl chloride (PVC) casing and sealed using epoxy. A stainless steel cathode is to be placed in the solution and connected to the negative output of a power supply, while the positive supply is connected to a stainless steel plate on the base of the specimen. A wet sponge will be placed between the stainless steel plate and the concrete to ensure connectivity.

A 13V potential difference will be applied so the negatively charged chloride ions, Cl⁻, will be forced down into the concrete. As chlorides penetrate the concrete, the resistivity of the concrete will change. Using an Ag/AgCl reference electrode, and a potentiostat connected to the two steel reinforcing bars, the macrocell current and half-cell potential of the steel will be monitored. When it is deemed the corrosion is active the system will be turned off.

During the migration of chlorides into the concrete the pond solution will become alkaline caused by leaching. After initiation of the corrosion, the specimen will be subjected to differing oxygen concentrations using three differing methods:

1) Seal the base of the concrete with wax preventing any further ingress of oxygen into the structure. The corrosion rate (i_{corr}) of the steel should therefore decrease with time if the corrosion is controlled by oxygen.

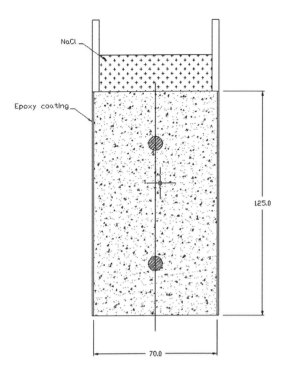

Fig. 3 General cross section of test method

2) Submerge the base of the concrete specimens in water to control the oxygen concentration by either bubbling oxygen to raise the concentration or nitrogen to lower it. This method can be used to compare the differing corrosion rates with changing oxygen concentrations as at present it is believed the more oxygen there is the faster the corrosion.
3) Subjecting the specimens to air can be used as a control specimen to compare the results from low oxygen concentrations.

4 Concluding Remarks

To determine service life estimation for submerged offshore structures, further understanding of the ageing of concrete and corrosion under low oxygen environments is required.

Test methods have been outlined and are being finalised to attempt to recreate the in situ conditions of a submerged structure where corrosion has initiated. Using controlled environments that limit the oxygen supply to the surfaces of the concrete specimens the corrosion rates will be monitored.

The research will lead to the development of a probabilistic model as an aid to determine the duration of offshore structures lifespan, which in turn will aid the decision making process in the decommissioning of these structures.

Research proposals are expected to take place over the following two years.

Acknowledgements Research is supported by Fairfield Energy Ltd, Atkins Oil and Gas and the Engineering and Physical Sciences Research Council.

References

[1] Concrete Offshore in the Nineties-COIN (1990), *A Summary Report*, HMSO Publication, OTH90 320.
[2] International Association of Oil & Gas Suppliers (2003), *Disposal of disused offshore concrete gravity platforms in the OSPAR Maritime Area*. International Association of Oil & Gas Suppliers, Report Number 338.
[3] OSPAR (1998), *OSPAR Decision 98/3 on the Disposal of Disused Offshore Installations*.
[4] North Sea Decomissioning Supply Chain Steering Group (2009), *Report on industry consultation*, Scottish Enterprise.
[5] Atkins Process Limited, Olav Olsen A/S (2003), *Decommissioning offshore concrete platforms*, HSE.
[6] CIRIA C674 (2010), *The use of concrete in maritime engineering - a guide to good practice*, CIRIA, London.
[7] Trethewey, K.R. (1988), *Corrosion: for students of science and engineering*, Longman Scientific & Technical, New York.
[8] Bertolini, L. (2004). *Corrosion of steel in concrete: prevention, diagnosis, repair*. Wiley-VCH, Weinheim.
[9] Böhni, H. (2005), *Corrosion in Reinforced Concrete Structures*, Boca Raton, Fla, CRC Press; Woodhead, Cambridge.
[10] Isgor, O.B. and Razaqpur, A.G. (2006), Can. J. Civil Eng., vol. 33, n. 6, p. 707.
[11] Raupach, M. (1996), Mater. Struct., vol. 29, n. 188, p. 226.
[12] Gjorv, O.E., Vennesland, O., El-Busaidy, A.H.S. (1986), Mater. Perform., vol. 25, n. 12., p. 39.
[13] Page, C.L. and Lambert, P. (1987), J. Mater. Sci., vol. 22, n. 3, p. 942.
[14] Yu, S.W. and Page, C.L. (1991), Cement Concrete Res., vol. 21, n. 4, p. 581.
[15] Hansson, C.M. (1993), Corros. Sci., vol. 35, n. 5, p. 1551.
[16] Castellote, M., Alonso, C., Andrade, C., Chadbourn, G.A., Page, C.L. (2001), Cement and Concrete Research, vol. 31, n. 4, p. 621.
[17] Gjorv, O.E., Vennesland, O. (1987), Cement Concrete Res., vol. 9, n. 2, p. 229.
[18] Thomas, M.D.A. and Bamforth, P.B. (1999), Modelling chloride diffusion in concrete: Effect of fly ash and slag. Cement Concrete Res., vol. 29, n. , p. 487.
[19] Castellote, M. and Andrade, C. (2006), Mater Struct., vol. 39, n. 10, p. 955.
[20] Lindvall, A. (2007), Cement Concrete Comp., vol 29, n. 2, p. 88.
[21] Castellote, M., Andrade, C., Alonso, C. (2001), Cement Concrete Res., vol. 31, n. 10, p. 1411.
[22] Wang, Y., Li, L-Y., Page, C.L. (2005), Build. Environ., vol. 40, n. 12, p. 1573.
[23] Han, S-H., Chae, J.W., Park, W-S., Yi, J-H. 2006. *Numerical modelling of deterioration in marine concrete structures*. In: Proc of the 16th int offshore and polar engineering conference, San Francisco, California, USA, May 28th - June 2nd 2006.
[24] Conciatori, D., Sadouki, H., Brühwiler, E. (2008), Cement Concrete Res., vol. 38, n. 12, p. 1401.
[25] Helland, S., Aarstein, R., Maage, M. (2010), Structural Concrete, vol. 11, n. 1, p. 15.
[26] Life-365 (2008), *Service Life Prediction Model and computer program for predicting the service life and life-cycle costs of reinforced concrete exposed to chlorides*, version 2.0 user manual.
[27] Visser, J.H.M, Gaal, G.C.M., Rooij, M.R. (2002), In: Proc of 3rd int RILEM workshop on testing and modelling the chloride ingress into concrete, Madrid, Spain, 9-10th September, p. 423.
[28] Oslakovic, I.S., Serdar, M., Bjegovic, D., Mikulic, D. (2008), In: Proc 11th int conference on durability of building materials and components, Istanbul, Turkey, 11-14th May, paper T11, p. 222.

[29] Nokken, M., Boddy, A., Hooton, R.D., Thomas, M.D.A. (2006), Cement Concrete Res. vol. 36, n. 1,p. 200.
[30] Stanish, K. and Thomas, M.D.A. (2003), Cement Concrete Res., vol. 33, n. 1, p. 55.
[31] Bamforth, P.B. (1999), Mag. Concrete Res., vol. 51, n. 2, p. 87.
[32] Djerbi, A., Bonnet, S., Khelidj, A., Baroghel-bouny, V. (2008), Cement Concrete Res., vol 38, n. 6, p. 877.
[33] Kato, E., Kato, Y., Uomoto, T. (2005), vol. 3, n. 1, p. 85.
[34] Ismail, M., Toumi, A., François, R., Gagné, R. (2008), Cement Concrete Res., vol. 38 n. 8–9., p. 1106.
[35] Castellote, M., Andrade, C., Alonso, C. (2002), Corros Sci, vol. 44, n. 11, p. 2409.

Electrically accelerated transport of chlorides in concrete considering non-linear chloride binding in non-equilibrium conditions

Przemek Spiesz and Jos Brouwers

Abstract The Rapid Chloride Migration test (NT Build 492) is commonly used in order to determine the chloride diffusion (migration) coefficient in concrete. This coefficient can then be used as the input value for service life design models for concrete structures in marine environment. However, the theoretical background of this test is still raising discussions as the equations describing the transport process of chlorides cannot predict correctly the chloride concentration profiles in concrete. The new, improved chloride transport model presented in this paper includes the non-linear nature of chloride binding in concrete as well as the non-equilibrium conditions between free- and bound- chloride concentrations, which were not taken into account in the traditional chloride transport model. This new model has been applied to the experimental data and the values of binding parameters, chloride mass transfer coefficient and the effective chloride diffusion coefficient were determined in the optimization process. The derived binding parameters (non-linear Freundlich isotherm) show very good agreement with values obtained from experimental measurements. The chloride mass transfer coefficient reveals a tendency to diminish during the RCM test. The effective diffusion coefficient determined based on the new chloride transport model is in line with the values found in literature.

1 Introduction

Chloride-initiated corrosion of the steel reinforcing concrete is the reason for deterioration of concrete elements and structures exposed to the seawater or de-icing salts in the majority of cases. In order to quantify the ingress speed of chlorides in concrete a number of laboratory test methods have been developed over years.

P. Spiesz (✉) • J. Brouwers
Eindhoven University of Technology, The Netherlands
e-mail: p.spiesz@tue.nl

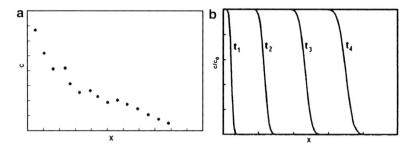

Fig. 1 Chloride concentration profiles after the RCM test: (a) experimental (b) theoretical

In the past, long-term diffusion methods were preferred, but nowadays the importance of short-term methods is increasing. One of very popular techniques used nowadays is the Rapid Chloride Migration test (RCM), developed in Sweden in the 90's (Tang [1]). The RCM test, standardized in the NT Build 492, is commonly used in order to determine the D_{RCM} - diffusion (migration) coefficient of chlorides in concrete. This diffusion coefficient can be further used as an input value for service life design models such as the DuraCrete model. Although the RCM test is very useful from the practical point of view (because of its short duration, simplicity of the procedure and non-sophisticated equipment required in order to execute the test), its theoretical background is not sound. As shown by few researchers [1–4], the chloride concentration profiles in concrete determined experimentally after the migration test (Figure 1a) are much different from the profile predicted from the mathematical model (Figure 1b) developed by Tang in [1].

The difference between the gradual experimental profiles and 'tsunami' shape theoretical profile gives evidence that the chloride transport model adopted for the RCM test is oversimplified, thus the D_{RCM} value calculated based on this model should be treated with scepticism.

2 New chloride transport model for the RCM test

The traditional model for the RCM test (Tang [1]) is based on the Nernst-Planck equation assuming (among other factors) that the binding of chlorides during the migration is linear and the equilibrium between free- and bound- chloride concentrations is achieved instantaneously. However, as commonly known, the binding of chlorides in concrete has a non-linear nature and the equilibrium between the free- and bound- chlorides cannot be achieved during the short-term migration test. As can be found in literature, the time which is needed to reach the equilibrium in concrete ranges from 7 up to 21 days, while the duration of the RCM test is usually limited to 24 hours. Based on these findings, a new chloride transport model which includes non-linear binding of chlorides (represented by Freundlich isotherm) in non-equilibrium conditions has been presented in this study.

2.1 Non-linear binding of chlorides in non-equilibrium conditions

The Freundlich isotherm is given by the following equation:

$$C_b = K_b \times c_s^n \qquad (1)$$

where: C_b – bo3und- chloride concentration, K_b – binding capacity, c_s – free- chloride concentration in the liquid at the liquid-solid interface and n – binding intensity parameter.

When chloride ions are transferred from one phase (liquid) to another (solid) across an interface that separates the two, the resistance to mass transfer causes a concentration gradient in each phase. Due to the limitations in the mass transfer through this interface, usually a certain time is required in order to achieve the equilibrium between the concentrations of chlorides in liquid and solid. The mass transfer rate of chlorides is proportional to the difference between the concentration of chlorides in the bulk solution and the equilibrium concentration at the liquid-solid interface, as given by the Freundlich isotherm (Eq. 1). Therefore, the mass transfer rate reads:

$$r = k(c - c_s) = k\left[c - \left(\frac{C_b}{K_b}\right)^{1/n}\right] \qquad (2)$$

where: r – mass transfer rate and k – mass transfer coefficient.

2.2 Formulation of the new chloride transport model in concrete for the RCM test

Applying the Nernst-Planck equation and Eq. (2), the following system of equations has been formed for non-steady state conditions, for free- and bound- chlorides respectively:

$$\varphi \frac{\partial c}{\partial t} + \frac{D_{eff} zFU}{RTL} \frac{\partial c}{\partial x} = -k\left[c - \left(\frac{C_b}{K_b}\right)^{1/n}\right] \qquad (3)$$

$$(1-\varphi)\rho_s \frac{\partial C_b}{\partial t} = k\left[c - \left(\frac{C_b}{K_b}\right)^{1/n}\right] \qquad (4)$$

where: φ – water-accessible porosity, t – rime, D_{eff} – effective chloride diffusion coefficient, z – ion valence, F – Faraday's constant, U – electrical voltage, R – universal gas constant, T – temperature, L – thickness of the specimen, x – distance and ρ_s – density of the solid state of concrete.

3 Application of the new chloride transport model

The system of equations (3) and (4) can be solved numerically. The forward discretization method has been applied in this study, resulting in the following equations for free- and bound- chlorides respectively:

$$c(i+1, j+1) = c(i+1, j) - \frac{\Delta x}{u}\left[\varphi \frac{c(i+1,j) - c(i,j)}{\Delta t} + k\left(c(i+1,j) - \left(\frac{C_{b(i+1,j)}}{K_b}\right)^{1/n}\right)\right] \quad (5)$$

$$C_{b(i+1,j)} = C_{b(i,j)} + \frac{\Delta t}{\rho_s(1-\varphi)} k\left(c(i,j) - \left(\frac{C_{b(i,j)}}{K_b}\right)^{1/n}\right) \quad (6)$$

with the initial and boundary conditions:

$$c(i, j=1) = c_0$$
$$C_{b(i=1,j)} = C_{bi} \quad (7)$$

where: i – time iteration, $i = 1 .. \, t/\Delta t$, j – distance iteration, $j = 1 .. \, L/\Delta x$, Δt – interval of time, Δx – interval of distance, u – chloride migration velocity ($u = D_{eff} zFU/RTL$), c_0 – concentration of chlorides in the bulk solution and C_{bi} – initial bound chloride concentration.

When all the experimental conditions, properties of concrete and the chloride concentration profile after the migration test are known, Eqs. (5) and (6) can be applied to this data in order to obtain the values of D_{eff}, k, K_b and n. In such an optimization process, the difference (the error) between the experimentally determined chloride concentration profile and the chloride profile computed based on Eqs. (5) and (6) is minimized by adjusting the values of above-mentioned parameters. In Table 1 the values of the optimized parameters are shown for the experimental data presented in [2]. In Figures 2 and 3 the measured and simulated total chloride concentration profiles for investigated concrete are shown as well.

As can be seen in Figures 2 and 3, the chloride concentration profiles simulated based on the new chloride transport model can predict the chloride transport in concrete due to migration much better in comparison to the traditional model (Figure 1b). The obtained binding parameters (K_b and n) are in line with chloride binding data found in literature for the same type of cement [5]. As can be observed in Table 1, the mass transfer coefficient (k), which is representing the non-equilibrium between free- and bound- chloride concentrations, is showing a tendency to diminish in time. This gives an indication that the chloride transport process between the pore solution and the active solid is becoming slower in time during application of electrical field. This effect needs a further investigation as it has not been studied yet. The optimized values of the effective chloride diffusion coefficient (D_{eff}) show a good agreement with values found in literature (e.g. Tang [1]) measured in steady-state tests for concretes with similar w/c ratio.

Table 1 Parameters optimized from application of the new model (experimental data derived from [2])

w/c	0.35	0.35	0.35	0.45	0.45	0.45
t [h]	6	9	18	6	9	18
D_{eff} [×10^{-12} m²/s]	0.85	1.08	0.96	2.04	2.15	1.92
k [×10^{-6} 1/s]	10.54	7.45	4.55	5.58	5.32	2.10
K_b [×10^{-3} dm^{3n}/g^n]	0.55	0.57	0.61	0.52	0.60	0.56
n	0.53	0.51	0.54	0.52	0.55	0.52

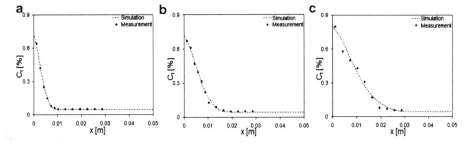

Fig. 2 Total chloride concentration profiles (C_t) obtained from the new model, concrete with w/c of 0.35, (**a**) t = 6 h, (**b**) t = 9 h, (**c**) t = 18 h

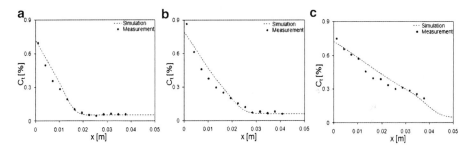

Fig. 3 Total chloride concentration profiles (C_t) obtained from the new model, concrete with w/c of 0.45, (**a**) t = 6 h, (**b**) t = 9 h, (**c**) t = 18 h

4 Conclusions

It is shown in this paper that the currently adopted theory for the Rapid Chloride Migration test does not predict the transport process of chlorides satisfactorily, which is reflected by the difference between the theoretical and experimental chloride concentration profiles.

In order to improve the migration model of chlorides in concrete, a non-linear chloride binding isotherm and non-equilibrium conditions between free- and bound-chlorides concentrations are implemented in the new model. The numerical solution of this new model has been applied to the experimental data. In the optimization process all the parameters have been adjusted in order to minimize the error between the simulated and measured chloride concentration profiles. The obtained values of the non-linear binding parameters (K_b and n) are showing a very good agreement with the values measured in speciation binding experiments of cement pastes. The mass-transfer resistance, represented by k, plays a decisive role on the chloride transport process, hence also on the shape of the chloride concentration profile in concrete. It is found in this paper that the value of k features a tendency to decrease in time during the migration test. This phenomenon has to be further analyzed, since no research has been carried out in this field yet. The effective diffusivity of chlorides is also derived in this paper by using the new theoretical model. This reveals a new possibility of application of the short-term non-steady-state migration test in order to obtain the effective chloride diffusivity, which is related only to the pore structure in concrete and thus is not influenced by binding. The D_{eff} obtained from the new model corresponds well with experimental data found in literature for similar types of concrete.

References

[1] Tang, L. (1996), *PhD Thesis*, Chalmers University of Technology.
[2] Stanish, K. D. (2002), *PhD Thesis*, University of Toronto.
[3] Yuan, Q. (2009), *PhD Thesis*, University of Ghent
[4] Gruyaert, E., Van den Heede, Ph., De Belie, N. (2009) in *Proceedings of the 2nd International RILEM Workshop on Concrete Durability and Service Life Planning*, pp. 440–448.
[5] Zibara, H. (2001), *PhD Thesis*, University of Toronto.

Chloride binding related to hydration products Part I: Ordinary Portland Cement

Miruna Marinescu and Jos Brouwers

Abstract A new hydration model of OPC [1, 2] was used in this study. The quantities of hydration products were calculated for saturated state hydration and for samples that have been dried to 11% rh after hydration. The chloride binding of several hydrated phases (C-S-H, monosulfate, hydroxy-AFm, AFt, CH, and Friedel's salt) has been studied and estimated in relation to the chloride content of the pore solution. Combining two new isotherms, for C-S-H and AFm chloride binding in hardened cement paste, and the computed quantities of all hydration products, a new formula for the chloride binding capacity of OPC pastes has been deduced. The new model is based on the chloride binding abilities of each hydrated phase, and therefore can distinguish between their contributions at different free chloride concentrations. When compared to experimental data, the results obtained using this new chloride binding isotherm have proven to accurately describe the chloride binding capacity of OPC pastes, results being within 12% of the experimental values for the whole range of free chloride concentrations considered, of up to 3 M.

1 The Chloride Binding Ability of the AFm Phase

Even though AFm is usually considered in cement literature to be only one phase, this generalization cannot be used when studying the microstructure of the cement paste. The AFm family consists of a great number of compounds, the most important being monosulfate (SO_4-AFm) $C_3A \cdot CaSO_4 \cdot 14H_2O$ and hydroxy-AFm (HO-AFm) $C_3A \cdot Ca(OH)_2 \cdot 12H_2O$. Moreover, intermediary compounds and solid solutions can form between these end-products.

The chloride ion is able to substitute the sulfate or hydroxyl groups in the structure of different AFm-type compounds, forming chloride-containing phases, Friedel's

M. Marinescu (✉) • J. Brouwers
Eindhoven University of Technology, The Netherlands
e-mail: m.marinescu@tue.nl

salt (3CaO·Al$_2$O$_3$·CaCl$_2$·10H$_2$O) and Kuzel salt (C$_3$A·1/2CaCl$_2$•1/2CaSO$_4$·10H$_2$O) among other possible compositions with a varying degree of anion substitution by the chloride ions. Different solid solutions can occur between these compounds and chloride-free AFm compounds. Despite the wide range of compositions of the AFm family, the difference in their chloride binding ability has never been detailed and quantified.

Hirao et al. [3] studied the chloride binding of an AFm phase which, given the synthesis route, can be assumed to be pure monosulfate. They have found the following Freundlich-type isotherm which describes chloride binding by the monosulfate phase, in mg Cl/ g monosulfate, as a function of the molar concentration of chlorides in the pore solution, termed c:

$$C^0_{b,SO_4-AFm} = 51.89c^{0.58} \qquad (1)$$

The pore solution is always considered to be in equilibrium with the external media, and therefore the same chloride concentration from the external solution is also used in these isotherms.

Birnin-Yauri and Glasser [4] consider that 100% of the quantity of HO-AFm is completely transformed into Friedel's salt at a free chlorides concentration greater than 0.015 mol/L. Considering the fact that one mole of hydroxy-AFm binds two moles of chloride ions in order to form Friedel's salt, the amount of chlorides which can be bound by C$_4$AH$_{13}$, in mg Cl/ g HO-AFm becomes:

$$C^0_{b,HO-AFm} = 126.5 \qquad (2)$$

This chloride binding capacity of HO-AFm is constant for external chloride concentrations over 0.015 M.

The sum of Eqns. (1) and (2) will be used henceforth to compute the total binding capacity of the AFm phase.

2 The Chloride Binding Ability of the C-S-H Phase

The surface of the C-S-H gel is known to be negatively charged, because the charges of the bridging silica thetraedra are not always compensated [7].Based on the interpretation of electrokinetic potential curves, a structure made of a condensed layer of Na$^+$ ions on the surface (Stern internal layer), compensated partly by SiO$^-$ groups and partly by an external layer (Stern external layer) made of hydroxyl ions (and chloride ions, in the presence of chlorides) was proposed. This external layer allows ionic exchanges with the solution. The adsorbtion of chloride ions can therefore be explained by an exchange mechanism between a chloride ion from the pore solution and a hydroxyl ion from the C-S-H layers. The OH$^-$ ion is loosely bound, permitting the Cl⁻ ion to be substituted in the interlayer space and to ensure the electroneutrality of the system. This physical adsorbtion mechanism of chloride ions on the surface of the C-S-H gel explains the fact that C-S-H has a lower chloride binding ability than the AFm phase, which binds chlorides through chemical substitution.

Eqn. (3) describes the chloride binding capacity of the C-S-H phase, computed in mg Cl/g C-S-H and termed $C^0_{b,C\text{-}S\text{-}H}$.

$$C^0_{b,C\text{-}S\text{-}H} = (6.65\, c^{0.334} \cdot \delta_{C_3S} + 7.89\, c^{0.136} \cdot \delta_{C_2S}) \quad (3)$$

in which δ_{C_3S} and δ_{C_2S} are the ratios of C_3S and respectively C_2S to their sum, which is the total amount of silicates in the cement mineralogical composition.

3 The Role of Ettringite, Portlandite and Friedel's Salt in Chloride Binding

In [8] it was concluded that, at intermediate levels of chloride concentrations (including 0.5 M NaCl), monosulphate is destroyed while most ettringite ($C_3A \cdot 3CaSO_4 \cdot 36H_2O$, the AFt phase) remains stable and that at high chloride levels both monosulphate and ettringite are destroyed with the release of Friedel's salt and gypsum as main final products. Birnin-Yauri and Glasser [4] propose that ettringite may play a role in chloride binding, even at lower free chloride concentrations, but this assumption is not supported by any experimental data. However, Hirao et al. [3] experimentally tested the chloride binding capacity of the AFt phase and found that ettringite does not bind any chloride ions from an external solution. Elakneswaran et al. [7] consider the chloride binding capacity of ettringite to be intermediary between the one of Friedel's salt and the one of C-S-H, which is a very low binding capacity. In this case, AFt is considered to bind chlorides not through a chemical process, but through physical adsorbtion on the hydrate's surface. Moreover, due to the small amounts of ettringite formed in the cement paste, when compared to the one of C-S-H, the effect of the chloride adsorption on the surface of ettringite would be minimal.

Elakneswaran et al. [7] also give chloride binding isotherms of portlandite (CH) and Friedel's salt, proposing that Friedel's salt can, in its turn, adsorb chloride ions on its surface. Portlandite is believed to behave in the same way. It is stated that the dissociation processes of both Friedel's salt and portlandite give positive surfaces. According to the same study, Friedel's salt dissociates in water and its surface charge (which is due to $[Ca_2Al(OH)_6]^+$ groups) is compensated by chloride ions in the solution. Chlorides can also be adsorbed on the positive surface of dissociated portlandite and form CaOHCl, its crystal structure being detectable by XRD. However, Hirao et al. [3] measured the chloride binding capacity of CH and found that it does not bind chloride ions.

4 Results and Conclusions

Based on an extensive analysis of experimental data found in literature [3, 5, 6], a new chloride binding isotherm for the AFm phase is given. This formula takes into account the different chloride binding abilities of monosulfate [3] and hydroxy-AFm [4]. This difference between the chloride binding abilities of various members of the AFm family has not been taken into consideration in any other model, even though it is significant and its inclusion can change the results greatly. Using the new AFm isotherm and several sets of experimental chloride binding isotherms from literature, a new formula for estimating the chloride adsorption capacity of the C-S-H phase, based on [5], is proposed.

Figure 1 shows isotherms $C^0_{b,SO_4\text{-}AFm}$ (Eqn. 1), $C^0_{b,HO\text{-}AFm}$ (Eqn. 2) and $C^0_{b,C\text{-}S\text{-}H}$ (Eqn. 3), describing the chloride binding ability of monosulfate, hydroxy-AFm and C-S-H, respectively, expressed in mg Cl/g hydrated phase. Figure 2 shows the same three isotherms, but applied to a hardened cement paste composition, so taking into account the proportion of each hydrated phase.

The contribution of HO-AFm is the greatest for all free chloride concentrations, but its importance decreases with the increase in free chlorides (from 60% at 0.3 M external chlorides, to 30% at 3M), as all HO-AFm is considered to be transformed to Friedel's salt, starting from external chloride concentrations of 0.015 M [4]. The contribution of monosulfate does not equal the one of HO-AFm until high chloride concentrations, but it increases with the increase in free chlorides, from 13 % at 0.3 M external chlorides concentration, to over 33% of the total chloride binding capacity.

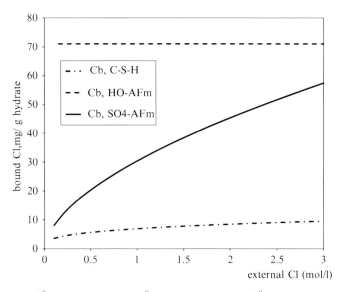

Fig. 1 Isotherms $C^0_{b,SO_4\text{-}AFm}$ (Eqn. 1), $C^0_{b,HO\text{-}AFm}$ (Eqn. 2) and $C^0_{b,C\text{-}S\text{-}H}$ (Eqn. 3) describing the chloride binding ability of hydroxy-AFm, monosulfate and C-S-H respectively, at 11% rh

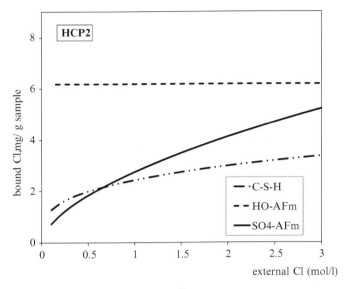

Fig. 2 Isotherm C_{b,SO_4-AFm}, $C_{b,HO-AFm}$ and $C_{b,C-S-H}$ applied to a hardened cement paste composition, taking into consideration the relative amount of each hydrate in the cement paste

The contribution of C-S-H remains fairly constant for the whole range of external chloride concentrations, between 25% and 28%. The lowest contributions are the ones of portlandite and Friedel's salt, whose sum amounts to only 2-5% of the total bound chlorides in an OPC paste. It can be concluded that, at lower concentrations, a high content in C_3A is beneficial for the ability of an OPC paste to bind chlorides, while at higher external chloride concentrations (for instance in salty lakes), the content in sulphates also becomes important.

5 Comparison with Other Models

The newly proposed model can be compared to two other models from literature [3, 5]. The highest drawback of the Zibara model [5] is that it is based only on a fit on cement composition and does not take into account other mixing and curing parameters. Therefore, this model predicts the same chloride binding capacity for pastes with different water/binder ratios, even though the authors have measured different values for each sample during the performed laboratory experiments. Applying the new hydration model [1, 2] to the model of Hirao et al. [3], the latter is found to perform well. In such conditions, the currently proposed model and the Hirao model [3] are compatible, and in good agreement with experiments. However, this hydration model assumes all AFm to consist of only one compound and does not make the distinction between the different chloride binding capacities of the AFm members, as the newly proposed model does.

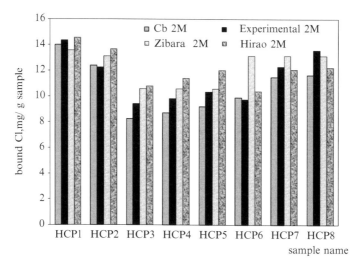

Fig. 3 The chloride binding capacity of different hydrated pastes with the experimental values at a free chloride concentration of 2 M

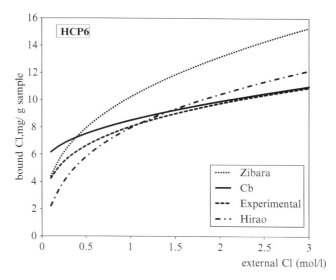

Fig. 4 Comparison between the proposed model, the models of [3] and [5] and experimental data from [5, 6]

Figure 3 shows a comparison between the new model, termed C_b, experimentally measured values from literature [5, 6] and the models of Zibara [5] and Hirao [3]. The chloride binding capacities of eight hydrated pastes with different OPC compositions, water/binder ratios and curing times are compared to the experimental values at a free chloride concentration of 2 M, as the regular value for the RCM test. Figure 4 is an example of the applied to one of the considered pastes [6] and it shows the isotherms obtained using the models of Zibara [5] and Hirao [3], comparing them to the experimental results and to the new chloride binding isotherm C_b. It can be seen from both figures that the proposed isotherm C_b is a very good fit for the considered experimental results.

References

[1] Brouwers, H.J.H. (2004), *Cem. Concr. Res.*, vol. 34, p. 1697.
[2] Brouwers, H.J.H. (2005), *Cem. Concr. Res.*, vol. 35, p. 1922.
[3] Hirao, H., Yamada, K., Takahashi, H., Zibara, H. (2005), *J. Adv. Concr. Tech.*, vol. 3, p. 77.
[4] Birnin-Yauri, U.A., Glasser, F.P (1998), *Cem. Concr. Res.*, vol. 28, p. 1713.
[5] Zibara, H. (2001), *Binding of External Chloride by Cement Pastes*, PhD Thesis, University of Toronto, Department of Building Materials.
[6] Tang, L., Nilsson, L. (1993), *Cem. Concr. Res.*, vol. 23, p. 247.
[7] Elakneswaran, Y., Nawa T., Kurumisawa K. (2009), *Cem. Concr. Res.*, vol.39, p. 340.
[8] Ekolu, S., Thomas, M., Hooton, R (2006), *Cem. Concr. Res.*, vol. 36, p. 688.

Chloride ingress in cracked concrete- a literature review

Branko Šavija and Erik Schlangen

Abstract Chloride induced corrosion of reinforcing steel is one of the most important mechanisms causing deterioration of reinforced concrete structures and the need for their premature repair or replacement. Significant research efforts have, therefore, been undertaken in recent decades, trying to quantify these effects. Yet, most of the studies and recommendations are based on the assumption of sound, uncracked concrete. However, reinforced concrete structures are frequently cracked, due to different causes, such as shrinkage, thermal effects, and loading. Cracking of the reinforced concrete structural members alters the local transport properties of the concrete cover, and allows rapid ingress of chloride ions and onset of corrosion. In the past two decades, several studies have focused on the influence of cracks on chloride ingress in concrete. This paper aims to review these research efforts, with respect to experimental methods used to produce cracked specimens, simulate harsh exposure conditions and analyze the results. Different influencing parameters are discussed, and some recommendations for further research are given.

1 Introduction

Reinforced concrete structures are commonly exposed to harsh environmental conditions. Nevertheless, major infrastructural works are often expected to have a service life of 100 years (sometimes even more), with little or no repair. In order to achieve this, structures must be durable. Therefore, service life prediction models which take deterioration mechanisms and material and mechanical properties of a structure into account are necessary. One of the major causes of deterioration of reinforced concrete structures is chloride ingress, which can cause corrosion of steel

---

B. Šavija (✉) • E. Schlangen
Microlab, Faculty of Civil Engineering and Geosciences,
Delft University of Technology, Delft, The Netherlands
e-mail: B.Savija@tudelft.nl

reinforcement. Structures exposed to marine environment or de-icing salts are, therefore, vulnerable. Since corrosion products have a larger volume than reinforcing steel, internal tensile stresses are imposed on the surrounding concrete, causing cracking, delamination and spalling of the concrete cover. Also, corrosion decreases the cross-sectional area of reinforcing steel bars and affects the bond between steel and concrete. Prediction of chloride ion penetration into concrete is, therefore, an important issue, both in research and in practise.

Many studies have been undertaken in order to quantify these effects, and several service life models have been proposed. Most of these studies have focused on determining transport properties of sound, uncracked concrete. However, reinforced concrete structures are designed to crack: in order for reinforcing steel to work efficiently, concrete must crack. This can significantly alter the transport properties of the concrete cover. In the past two decades several researchers have investigated the effect of cracking on chloride ion penetration in concrete, and its consequences [2–24, 27, 30, 32, 34]. When the effect of cracking on corrosion of reinforcing steel is concerned, there are two schools of thought: while some researchers consider cracking to accelerate both the onset of corrosion and its propagation [17], others argue that, even though cracks could shorten the initiation phase, the propagation of corrosion is not affected [6, 20]. Some researchers also claim that, by defining a service life of a structure as the initiation phase, engineers grossly underestimate the actual service life of a cracked concrete structure. The reported findings of a 17-year long study by François et al. [32] suggest that the propagation phase could be longer than the initiation phase. According to these authors, it is necessary to model the propagation phase and to be able to define the end of serviceability. In any case, it is essential to conduct further research on the influence of cracks on both the initiation and propagation phases.

In the following chapters, the emphasis is on the different methods used by various researchers to study the transport of chloride ions in cracked concrete, and their attempts to quantify different influencing parameters. The corrosion behaviour of steel in cracked concrete is not the primary aim of this paper, and is therefore only briefly mentioned.

2 Specimen preparation

In order to study the effects of cracking on transport properties of concrete, cracked specimens first need to be prepared in the laboratory. Researchers have used a variety of different methods to induce cracking. Most of the studies have aimed at producing cracks of controlled widths. The cracks which are produced can be realistic (using destructive techniques) or artificial (different non-destructive techniques). Some authors have also produced cracks of widths which were not controlled, for example by freezing and thawing [10], or by exposing the specimens to hygral and thermal cycles [24]. Tests on specimens with cracks of controlled widths aim to investigate the influence of cracks of different magnitudes on transport properties of concrete, and to determine whether the allowed values prescribed by current codes of practice should be changed. Therefore, these methods are further discussed.

The main destructive methods used by researchers are:

Splitting method. Splitting test (Brazilian splitting test) is often used to measure tensile strength of concrete. It is an easy test to do, and its modification was therefore adopted by several researchers in their study [3, 5, 27–29]. The test is carried out on cylindrical or disc samples. Specimens are compressed in diametrical direction, usually through plywood strips placed between the specimen and the loading machine, to prevent crushing at the contact points. In this way, tensile stresses are imposed in the specimen, and it cracks in the direction of the loading. In order to obtain width controlled cracks, feedback-controlled tests are used. Two LVDT (linear variable displacement transducer) sensors, one on each side of the specimen, are used to measure the displacement normal to the axis of the loading, and their average (called COD- crack opening displacement) is used to control the load. After the target COD is obtained, the specimen is unloaded, and the crack partially closes. It is, therefore, necessary to measure the crack width in unloaded state as well. However, as stated by the authors of the study [29], cracks produced in this way may be identical in widths on the inside as on the outside, while the crack widths adopted by the codes are based on flexure induced (i.e. V shaped) cracks. Specimens are exposed to chlorides in non-loaded state. These specimens are, typically, not reinforced.

Bending method. The test is carried out on prismatic beam specimens [2, 6, 7, 13, 15, 17, 19-21, 32]. Depending on the setup, a force is applied in the middle of the beam (3 point bending), or in the thirds of the beam (4 point bending). The load is applied until bending cracks appear. Crack width can be controlled by using LVDT sensors to monitor and control the CMO (crack mouth opening). Alternatively, the load needed to produce a certain crack width can be estimated by calculation, and then applied [2]. Cracked beams are sometimes loaded back-to-back in order to keep the crack open during the chloride exposure. Single crack and multi crack specimens can be obtained. Beams can be exposed to chlorides in loaded condition. Specimens are typically reinforced.

Wedge splitting method. This method is commonly used in fracture mechanics, because it is very simple. A prismatic specimen is provided with a groove and a notch. The wedge is installed in the groove and loaded in compression in the vertical direction. Rollers turn the compressive vertical load into two horizontal loads, which move away from each other and crack the starter notch. Using LVDTs, the CMO can be measured, and the width of the crack can be controlled. This method was utilized in a study by Pease [34]. A modification of this method was used in the study by Yoon et al. [22]. Cylindrical specimens were used. First, steel plates were glued to a notch, and then loaded in tension. After the target CMO was achieved, specimen was unloaded. Effective CMO (after unloading) was measured and recorded. Cracks created in this manner are V-shaped cracks, and can be considered realistic. Unloaded specimens are exposed to chlorides. These specimens are not reinforced.

Expansive core method. This method was used in studies by Ismail et al. [8, 9] to produce controlled cracks in brick and mortar specimens. The specimens used were doughnut shaped discs (50 mm height, internal diameter 50 mm, external diameter 100 mm). They were loaded using a mechanical expansive core and external steel confinement ring. Deformations imposed on the internal diameter of the specimens by the expansive core and the confinement pressure applied by the external steel ring can be adjusted, and cracks with selected openings produced. Cracks created this way have a different shape when compared to flexure induced cracks (perhaps similar to cracks created using the splitting method, i.e. almost parallel wall cracks), and are not, therefore, directly comparable to those considered in structural design codes. While loaded, these specimens were fitted in the chloride penetration cell. Specimens used in these studies were not reinforced.

Several studies were conducted on specimens with artificial cracks. In the study [18], artificial smooth cracks were produced by saw cutting of concrete cylinders longitudinally into two equal width parts. The cracks were created by clamping the cut cylinder parts back together and using brass shims of various thicknesses at the edges to keep the gap open. In studies [4, 14] artificial cracks were produced by placing copper sheets in a hardening concrete specimens upon casting, and removing them after approximately 4 hours. Different copper sheets were used to obtain cracks of different widths and depths. Artificial cracks are, in general, parallel walled, and their shape does not resemble the shape of flexure induced (i.e. tapered) cracks. While these artificial cracks can be used to study the influence of different parameters affecting the chloride ingress, they also lack the tortuosity, connectivity, and roughness of real cracks. It is also possible that transport properties are altered by the wall effect in the cracks (studies [4, 14]), and that lack of contact between the surfaces prevents possible autogeneous healing of cracks, which is an important mitigating mechanism for chloride transport. Nevertheless, this method has proven to be a simple and effective tool to study influence of different crack depths and widths on chloride ingress.

It can be concluded that specimens cracked using bending method give the most realistic results: cracks are V shaped, tortuous, with rough surfaces, and the specimens can be kept loaded during exposure to chlorides. Also, these specimens are usually reinforced, so corrosion measurements can also be performed on them. Crack blocking by corrosion products [13] can also be studied using reinforced specimens. Other methods have, however, numerous other advantages, (simplicity, for example) and can be, therefore, successfully used to study different influencing parameters.

3 Methods used to test chloride ingress in cracked concrete

Chloride transport in concrete is governed by several different mechanisms: capillary absorption, hydrostatic pressure, diffusion, and migration. Methods used to test chloride ingress in concrete mainly assume that either diffusion or migration is a governing mechanism for transport of chloride ions. However, some authors [21] suggest that capillary suction, when it takes place, has an influence which is greater than the diffusion mechanism, and therefore should not be ignored for real structures. Several testing methods have been developed for testing chloride transport in sound concrete. Some of these methods have been used by researchers to assess the effect of cracking on chloride ion penetration, and these are briefly addressed. More details about testing the chloride penetration resistance of (sound) concrete and detailed descriptions of the methods can be found elsewhere, for example [33].

3.1 Diffusion tests

Chloride diffusion into concrete, like any other diffusion process, is controlled by Fick's laws (Fick's First Law for steady state diffusion, and Fick's Second Law for non-steady state diffusion). Even though diffusion tests are similar to natural conditions, their major flaw is their long duration.

Several researchers have used immersion in salt water solution to study the effects of cracking on chloride ingress [2, 7, 13, 18, 21, 23]. The cracked specimen is immersed in artificial salt water, and left for a long time period, before the chloride profiles are obtained. In the study by Adiyastuti [2], cracked beams were immersed in 3% NaCl solution for up to 2 years. Yoon and Schlangen [23] immersed the cracked specimens in artificial chloride solution with NaCl and $MgCl_2$, and compared the results with those of a migration (short term) test. A standardised immersion method is, for example, NordTest NT Build 443 [25], which was used in the study by Garces Rodriguez and Hooton [18]. Ismail et al. [8, 9] used steady-state diffusion test to study the effects of cracking. They fit the cracked specimen into a diffusion cell, and used it to determine the diffusion coefficients. Another method is to expose cracked specimens to cyclic wetting and drying (non-steady state diffusion) [6, 15–17, 21, 32]. François et al. [6, 32] constructed a confined chamber in which they placed cracked beam specimens. The specimens were exposed to periodic wetting and drying by spraying salt fog from the corners of the chamber.

3.2 Electrical migration tests

Chloride penetration is sometimes accelerated by applying an electrical field as a driving force to move chloride ions into the exposed concrete. The movement of ions in a solution under an electrical field is governed by the Nernst-Planck equation. This is a fast method to determine the chloride ion transport in concrete. However, it is not directly applicable to real structures, and caution should be taken in interpreting the results.

NordTest NT Build 492 [26] is a standardised non-steady state migration test. Even though it was meant to be used on sound concrete specimens, several authors have used the modified version of the test to asses the effects of cracks [4, 14, 22]. Yoon et al. [22] used this method to quantify the effects of different crack widths on chloride penetration. Audenaert et al. [4] and Marsavina et al. [14] used it to assess the penetration resistance of concrete with different artificial cracks. Apart from this method, some studies utilized the steady state migration tests to study cracking. Djerbi et al. [5] used this method to assess migration of chloride ions in specimens made using two different concrete mixtures, with varying crack widths.

Other test methods exist (for example, electricity conductivity test- ASTM C1202), which are used to assess concrete's ability to resist chloride ion penetration, but these are often criticized and not so widely used in the literature.

4 Factors influencing chloride penetration in cracked concrete

A lot of parameters can influence chloride penetration, for example: crack width, crack depth, binder type, binder content, water-to-binder ratio, loading and exposure conditions, exposure duration, etc. Several mitigating mechanisms can, possibly, slow down the ingress of chloride ions in cracks, namely autogeneous (self) healing, and blocking of cracks by corrosion products. While influence of some factors is thoroughly examined (exposure condition, exposure duration) in studies on sound concrete, similar results are reported for cracked concrete. For example, Win et al. [21], in a study which used different concentrations of chloride solutions (3-8% NaCl solution), reported that higher NaCl concentrations lead to deeper penetration and higher chloride content. Also, it is reported that the longer test duration leads to deeper chloride penetration [2, 4, 14, 21]. However, some influencing factors are only present when cracked specimens are examined. These factors are further discussed.

4.1 Crack width

Crack width is considered to be one of the most important parameters when assessing the durability of a concrete structure. Most of the design codes limit the (surface) crack width in ranges from 0.15 to 0.30 mm. However, the validity of these limit values has yet to be proven.

Djerbi et al. [5] found that the diffusion coefficient of cracked concrete increased with the increasing crack width and was almost constant for cracks wider than 0.08 mm. Similar findings were reported by Ismail et al. [9]: for a crack wider than 0.2 mm, the diffusion of chloride ion in the solution is not a limiting factor controlling the diffusion process perpendicular to the crack wall; for 0.08-0.1 mm cracks, diffusion process still occurs, but it is much slower; for cracks smaller than 0.03 mm, no diffusion occurs along the crack path. Yoon and Schlangen [23] reported the following: for short term (rapid) test, cracks smaller than 0.012 mm have no influence on penetration depth; however, in a long-term test, the threshold width was found to be 0.05 mm. They have attributed this difference to crack healing. On the contrary, Marsavina et al. [14] found no pronounced influence of crack width on chloride penetration in their study. However, crack widths used in this study were in the range 0.2-0.5 mm, therefore larger than the limit values proposed in the other studies, so this is in agreement with the other results.

4.2 Crack depth

In the study by Marsavina et al. [14] it was concluded that crack depth has a more pronounced influence on chloride ion penetration than crack width. Same results were reported by Audenaert et al. [4], who concluded that this effect is more pronounced for longer test durations. Both of these studies utilized the artificial crack method.

4.3 Water-to-binder ratio, binder content, binder type

Win et al. [21] studied the effect of different w/c ratios on chloride penetration ordinary Portland cement cracked concrete. They concluded that the specimens with low w/c ratio (0.25) showed lower concentration profile and penetration depth both from exposed surface and around the crack (compared with w/c ratios of 0.45 and 0.65). The increase in w/c led to a higher ingress rate of ions, not only from the exposed surface, but also around the crack.

Audenaert et al. [4] found, in their study, that increasing the amount of cement, at constant w/c ratio, led to a decrease in chloride penetration depths. Studies [11, 16, 18] all reported a decrease in penetration depths when using blended cements or supplementary cementitious materials (slag, silica fume). Both of these effects (i.e. decreased chloride penetration with an increase in cement content and with use of blended cements) could be attributed to increased chloride binding under these conditions. Further study on the effects of blended binders on chloride penetration in cracked concrete is needed.

4.4 Loading conditions

Even though most of the tests are done on unloaded specimens, it is important to take into account that in reality all of the structural members are mechanically loaded. Gowripalan et al. [7] investigated the influence of flexural loading on chloride penetration. They concluded that the apparent diffusion coefficient is higher in the tension than in the compression zone of the beam element, and attributed this to the damage at the aggregate-paste interface in the tension zone which can expedite diffusion, while in the compression zone a reduction of porosity is present, which can slow down the diffusion. Küter et al. [30] studied the effect of cyclic loading on chloride ingress in cracks, and concluded that it is more severe than in the case of static loading conditions. It is, therefore, important to take realistic loading conditions into account when assessing the service life of a structure. Further research in this area is also necessary.

4.5 *Mitigating mechanisms*

It is known that concrete possesses a certain ability to heal cracks under certain conditions. This mechanism is called autogeneous (self) healing of concrete [31]. Some authors found that it is an important factor that needs to be taken into account when studying chloride penetration in cracked concrete [6, 9, 10, 16, 19, 23]. According to a study by Jacobsen et al. [10], self-healing of cracked concrete specimens for three months in water led to a significant reduction in rate of chloride migration: 28-35%, compared to migration in newly cracked specimens. Sahmaran [19] stated that formation of self-healing products slows down the rate of chloride penetration through the crack zone of mortar. This mechanism needs to be, therefore, further studied, especially with regard to different binder types.

Another mitigating mechanism which has been reported is crack blocking by corrosion products. Marcotte and Hansson [13] found, in their study, that crack blocking by corrosion products occurred in concrete made using 10% silica fume replacement. This phenomenon is also mentioned elsewhere in the literature, although not widely studied. Further research is needed.

5 Conclusion

Reinforcement corrosion is a serious problem in modern concrete structures. Even though significant research efforts have been undertaken in recent decades, most of the studies were focused on sound, uncracked concrete. These studies provided us with significant insight on the involved phenomena. Still, the findings cannot be applied in some practical situations, where concrete cracking is encountered or expected. It is, therefore, important to study the influence of cracking on chloride transport in concrete. Different methods were utilized by authors to create cracks and expose the specimens to chlorides. A lot of influencing factors have been examined, and still the consensus has not been reached on the importance and magnitude of most of them. Even though structural design codes prescribe maximum allowed (surface) crack widths in aggressive environments, it is still uncertain whether these values have a sound scientific background or not. This demands further research. Also, as the construction industry is becoming more and more environmentally conscious, the effects of using more eco-friendly, blended binders (fly ash, blast furnace slag) on chloride penetration in cracked concrete should be thoroughly examined. The effects of mechanisms which could potentially slow down chloride penetration (such as autogeneous (self) healing and crack blocking) should not be neglected in further studies and, eventually, service life models. Since this area of research has been getting a lot of attention lately, it is justified to believe that increased understanding of deterioration mechanisms of cracked concrete will, eventually, lead to practical service life models for real, cracked structures.

Acknowledgement Financial support by the Dutch Technology Foundation (STW) is gratefully acknowledged.

References

[1] Neville A. (1995), *Mater Struct*, vol. 28, pp. 63–70
[2] Adiyastuti, S.M. (2005), *Influence of cracks on chloride induced corrosion in reinforced concrete flexural members*, PhD Thesis, University of New South Wales, Sydney, Australia
[3] Aldea, C.M., Shah, S.P. and Karr, A. (1999), *J Mater Civil Eng*, vol. 11, n.3, pp. 181–187
[4] Audenaert, K., Marsavina, L. and De Schutter, G. (2009), *Key Eng Mat*, vol. 399, pp. 153–160
[5] Djerbi, A., Bonnet, S., Khelidj, A. and Baroghel-bouny, V. (2008), Cement Concrete Res, vol. 38, pp. 877–883
[6] François, R. and Arliguie, G. (1998), *J Mater Civil Eng*, vol. 10, n. 1, pp. 14–20
[7] Gowripalan, N., Sirivivnaton, V. and Lim, C.C. (2000), *Cement Concrete Res*, vol. 30, pp. 725–730
[8] Ismail, M., Tuomi, A., François, R. and Gagne, R. (2004), *Cement Concrete Res*, vol. 34, pp. 711–716
[9] Ismail, M., Tuomi, A., François, R. and Gagne, R. (2008), *Cement Concrete Res*, vol. 38, pp. 1106–1111
[10] Jacobsen, S., Marchand, J. and Boisvert, L. (1996), *Cement Concrete Res*, vol. 36, pp. 869–881
[11] Konin, A., François, R. and Arliguie, G. (1998), *Mater Struct*, vol. 31, pp. 310-316
[12] Lim, C.C., Gowripalan, N., and Sirivivatnon, V. (2000), *Cement Concrete Comp*, vol.22, pp. 353–360
[13] Marcotte, T.D., Hansson, C.M. (2003), *J Mater Sci*, vol. 38, pp. 4765–4776
[14] Marsavina, L., Audenaert, K., De Schutter, G., Faur, N. and Marsavina, D. (2009), *Constr Build Mater*, vol. 23, pp. 264–274
[15] Mohammed, T.U., Otsuki, N., Hisada, M. and Shibata, T. (2001), *J Mater Civil Eng*, vol. 13, n.3, pp. 194–201
[16] Otieno, M.B., Alexander, M.G. and Beushausen, H.-D. (2010), *Mag Concrete Res*, vol. 62, n.6, pp. 393–404
[17] Otsuki, N., Miyazato, S., Diola, N.B. and Suzuki, H. (2000), *ACI Mater J*, vol. 97, n. 4, pp. 454–464
[18] Garces Rodriguez, O. and Hooton, R.D. (2003), *ACI Mater J*, vol. 100, n.2, pp. 120–126
[19] Sahmaran, M. (2007), *J Mater Sci*, vol. 42, pp. 9131–9136
[20] Schießl, P. and Raupach, M. (1997), *ACI Struct J*, vol. 94, n. 1, pp. 56–61
[21] Win, P. P., Watanabe, M. and Machida, A. (2004), *Cement Concrete Res*, vol. 34, p. 1073–1079
[22] Yoon, I. S., Schlangen, E., de Rooij, M.R. and van Breugel, K. (2007), *Key Eng Mat*, vols. 348-349, pp. 769–772

[23] Yoon, I. S. and Schlangen, E. (2010), *Key Eng Mat*, vols. 417-418, pp. 765–768
[24] Taheri-Motlagh, A. (1998), *Durability of reinforced concrete structures in aggressive marine environment*, PhD Thesis, Delft University of Technology, Delft, The Netherlands
[25] NordTest NT BUILD 443 (1995), Finland
[26] NordTest NT BUILD 492 (1999), Finland
[27] Jang, S. Y., Kim, B.S., and Oh, B.H. (2011), *Cement Concrete Res*, vol. 41, pp. 9–19
[28] Aldea, C.M., Shah, S.P. and Karr, A. (1999), *Mater Struct*, vol. 32, pp.370–376
[29] ang, K., Jansen, D.C. and Shah, S.P. (1997), *Cement Concrete Res*, vol. 37, pp.381–93
[30] Küter, A., Geiker, M.R., Olesen, J.F., Stang, H., Dauberschmidt, C. and Raupach, M. (2005). In: Proceedings of ConMat '05, Vancouver, Canada
[31] Schlangen, E. and Joseph, C. (2008) In: *Self-Healing Materials: Fundamentals, Design Strategies, and Applications*, Chapter 5, pp. 141-182, Ghosh S. K. (Ed.), WILEY-VCH Verlag GmbH & Co., Weinheim
[32] François, R., Castel, A., Vidal, T. and Vu, -N.A. (2006), J Phys IV, 136, pp.285-293
[33] Stanish, K.D., Hooton, R.D. and Thomas, M.D.A. (1997), FHMA Contract DTFH61, Department of Civil Engineering, University of Toronto, Canada
[34] Pease, B.J. (2010), *Influence of concrete cracking on ingress and reinforcement corrosion*, PhD thesis, Technical University of Denmark, Lyngby, Denmark

Numerical simulation of reinforcement corrosion and protection in submerged hollow concrete structures

Andrea Della Pergola, Federica Lollini, Elena Redaelli, and Luca Bertolini

Abstract Generally the corrosion conditions of steel in the submerged zone of reinforced concrete structures exposed to seawater are not severe, due to the low oxygen content of saturated concrete, the low steel potential and the high value of the chloride threshold level. However, in hollow structures macrocells could develop, which can increase the steel potential and corrosion rate. For this reason, in these structures cathodic prevention or cathodic protection systems could be useful. For several reasons, it is difficult to study the effects of macrocells and the effectiveness of cathodic protection systems in real structures. This paper describes results of numerical models aimed at simulating corrosion and protection of steel reinforcement in hollow structures.

1 Introduction

Chloride-induced corrosion is the main cause of degradation of reinforced concrete structures exposed to marine environments. According to Tuutti's model, steel corrosion develops in two phases: in the first phase, called initiation period, steel is embedded in alkaline and chloride-free concrete. In this condition steel is protected by a passive film and its corrosion rate is negligible. However, when structures are exposed to chloride environments, during the initiation period chlorides can penetrate through the concrete cover and reach the steel surface. When chlorides concentration on steel surface reaches a critical level, called chloride threshold level (CTL), steel depassivation occurs and corrosion can propagate.

The value of the CTL depends on many factors, e.g. concrete composition, steel composition, role of the steel-concrete interfacial zone, steel surface treatments, environmental exposure. A key factor is the steel potential. The relationship between

---

A.D. Pergola (✉) · F. Lollini · E. Redaelli · L. Bertolini
Politecnico di Milano, Dipartimento di Chimica, Materiali e Ingegneria Chimica, Milan, Italy
e-mail: andrea.dellapergola@mail.polimi.it

CTL and steel potential was initially studied by Pedeferri [1] who showed that when the steel potential is low, depassivation occurs at a much higher chloride content [1]. This was also confirmed by other authors [2].

According to the Pedeferri diagram, usually in submerged concrete structures a high chloride content is needed to cause steel depassivation, because the steel potential is low due to the continuous lack of oxygen. However, hollow structures are an exception, because in these structures the external concrete surface is in contact with seawater, but the internal surface is in contact with air. In this type of structures macrocells can develop, which may lead to an increase of steel potential and corrosion rate of steel in the saturated concrete. Hence, there is the need of studying the effect of macrocells between bars respectively in water-saturated and aerated concrete, as well as to study protection systems such as cathodic protection or cathodic prevention [3].

For several reasons, it would be rather complex and time-consuming to study steel corrosion conditions and steel protection on real submerged structures. Indeed corrosion monitoring systems commonly used in the atmospheric zone do not work properly when installed in the submerged zone of structures [4]. The detection of the effects of macrocell is rather difficult and, moreover, it would be necessary to spend a long time to get a regime humidity profile and chloride profile. Numerical models may be useful to study steel corrosion conditions under the effect of macrocells and to study the effectiveness of cathodic protection and cathodic prevention systems.

This paper concerns Andrea Della Pergola's PhD project, which is aimed at studying the role of preventative techniques against chloride-induced corrosion. The investigation of the CTL in the presence of preventative techniques (such as the use of stainless steel or cathodic prevention systems) and the development of a methodology for the evaluation of the CTL when it reaches high values are the main goals. Within this project corrosion and protection of steel in submerged hollow structures has been studied by numerical models. The preliminary results of this part will be presented in the following.

2 Models

A commercial finite element program was used to simulate a portion of reinforced concrete wall with one side exposed to seawater and the other side exposed to air. The program calculates potential and current density on steel surface, integrating Laplace equation [5–12].

2.1 Geometry

A wall 640 mm thick with longitudinal reinforcement both in the outside part exposed to seawater and in the inner part exposed to air was considered. The centre-to-centre distance between the bars of the two layers was 500 mm, while the centre-to-centre distance between two bars of the same layer was 150 mm. The bar

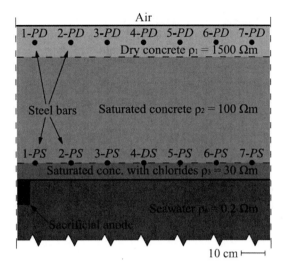

Fig. 1 Geometry of the model with a depassivated steel bar in saturated concrete (bar number 4) and a sacrificial anode placed in seawater.

diameter was 20 mm and the concrete cover on each side of the wall was 60 mm thick. Both a configuration without sacrificial anodes and a model with prismatic sacrificial anodes with a 100 mm x 100 mm cross section placed every 2.10 m were taken into account. Taking advantage of the axis of symmetry of the geometry, only a 1.05 m wide portion of the wall was considered (Figure 1). To keep the model geometry as simple as possible and to limit the calculation time needed by the program, two-dimensional models were created: in these models only longitudinal elements could be represented, while transversal elements (such as stirrups) had to be neglected. 3D models will be also considered to validate the results of 2D models.

2.2 Subdomain conditions

As subdomain conditions, electrical resistivity for each material was chosen. Four subdomains were considered (Figure 1): 1) dry concrete in contact with air, with thickness of 130 mm and constant resistivity of 1500 $\Omega \cdot$m; 2) saturated concrete in the bulk of the wall, with thickness of 440 mm and constant resistivity of 100 $\Omega \cdot$m; 3) concrete saturated by seawater, with thickness of 70 mm and constant resistivity of 30 $\Omega \cdot$m; 4) seawater, with thickness of 2000 mm and constant resistivity of 0.2 $\Omega \cdot$m.

2.3 Boundary conditions

Different types of boundary conditions were defined. On the perimeter of the model electric insulation was applied, considering that current does not flow through those surfaces. On the perimeter of the anode (in the model where the anode was present) a condition of constant potential was applied, because potential of an anode placed in seawater only slightly depends on the supplied current (E_{ANODE} = -1050 mV/SCE). On the steel surface, polarization curves were applied, that describe the relationship between current density i and steel potential E. The behaviour of active steel was described by Eqn. (1), the Butler – Volmer equation:

$$i = i_A + i_C = i_{CORR} \cdot \left(10^{\frac{E-E_{CORR}}{b_A}} - 10^{\frac{-(E-E_{CORR})}{b_C}} \right) \quad (1)$$

where i (mA/m²) is the net current density on steel surface; i_A and i_C (mA/m²) are, respectively, the anodic and cathodic current densities; i_{CORR} (mA/m²) is the corrosion current density; E (mV/SCE) is steel potential; E_{CORR} (mV/SCE) is free corrosion potential; b_A and b_C (mV/dec) are the anodic and cathodic Tafel slopes. The behaviour of passive steel was described by Eqn. (2) that takes into account the limiting current density due to restricted oxygen diffusion:

$$i = 1 - 10^{\frac{-(E-E_{CORR})}{b_C}} \bigg/ \left(\frac{1}{i_{PAS}} + \frac{10^{\frac{-(E-E_{CORR})}{b_C}}}{i_{LIM}} \right) \quad (2)$$

where i_{PAS} (mA/m²) is the anodic passive current density; i_{LIM} (mA/m²) is the cathodic limiting current density. The values of all constants of Eqn. (1) and (2) are listed in Table I. Figure 2 shows the polarization curves of passive steel in dry concrete (*PD*); passive steel in saturated concrete (*PS*); depassivated steel in saturated concrete (*DS*).

Different phases of the lifetime of the structure were simulated. Phase 1 corresponds to the initiation period of corrosion: in this phase all steel bars, both in dry concrete and in saturated concrete, were still passive, so the *PD* and the *PS* curves were applied as boundary conditions; phase 2 represented the propagation period: in this phase steel depassivation occurred on one steel bar in chloride-contaminated concrete, so the *DS* polarization curve was applied on bar number 4 in Figure 1.

The results of four models will be presented. Two models refer to structures not protected with sacrificial anodes: during the initiation period (phase 1) and after steel depassivation (phase 2). Other models refer to structures in which steel depassivation is prevented providing sacrificial anodes (cathodic prevention) or corrosion rate is controlled installing sacrificial anodes after steel depassivation (cathodic protection).

Table I Parameters used for the polarization curves of passive steel in dry concrete (*PD*), passive steel in saturated concrete (*PS*) and depassivated steel in saturated concrete (*DS*)

	E_{CORR} (mV/SCE)	i_{PAS} (mA/m²)	i_{CORR} (mA/m²)	i_{LIM} (mA/m²)	b_A (mV/dec)	b_C (mV/dec)
PD	-100	0.1	-	1000	-	300
PS	-400	0.1	-	0.2	-	300
DS	-800	-	0.2	-	75	10000

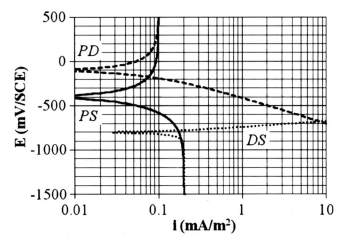

Fig. 2 Polarization curves of steel bars, depending on their exposure conditions

3 Results and discussion

3.1 Wall without sacrificial anode

Results obtained from the model without a sacrificial anode showed that in submerged hollow structures macrocell formation had an important influence both on steel corrosion initiation (phase 1) and propagation (phase 2). Indeed during the initiation period, when both steel bars in saturated and in dry concrete were still passive, a macrocell develops among these bars because of their different corrosion potential. In this phase the corrosion rate of passive steel in saturated concrete was limited by the value of passivity current density, so it could not exceed 0.1 mA/m^2. However, this simulation showed that passive steel bars in saturated concrete were subject to a sharp anodic polarization (exceeding 200 mV). Figure 3 shows steel potential of passive bars in saturated concrete in free corrosion condition (black triangles) and under the influence of macrocell (white triangles): the difference between these values is the anodic polarization. Figure 3 also shows the free corrosion potential (black circles) and the macrocell potential (white circles) of passive steel in dry concrete. Because of steel passivity conditions, the macrocell was under anodic control and the macrocell current circulating in the domain was very low. For this reason the ohmic drop between the anodic and the cathodic zone was negligible: that is why in Figure 3 the potential of steel bars in saturated concrete and in dry concrete are overlapped during the macrocell. The increase of about 200 mV in the steel potential of the passive bars in water saturated concrete, according to Pedeferri diagram, leads to a remarkable decrease of the chloride threshold level and thus may promote corrosion initiation when chlorides penetrate the concrete cover.

Fig. 3 Potential of steel bars in phase 1: in saturated concrete (symbol Δ) and in dry concrete (symbol O) in free corrosion conditions (black symbols) and under the influence of macrocell (white symbols)

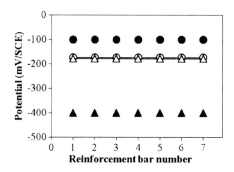

Fig. 4 Potential of steel bars in phase 2: in saturated concrete (symbol Δ) and in dry concrete (symbol O) in free corrosion conditions (black symbols) and under the influence of macrocell (white symbols)

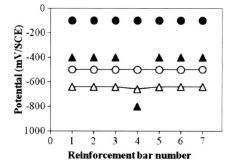

Once corrosion has initiated a different macrocell develops. Figure 4 and Figure 5 show the results of a simulation of the macrocell that forms when corrosion initiates on the middle of the steel bars in saturated concrete. A macrocell developed among this bar, which acted as an anode, and all the other passive ones, both in saturated and in dry concrete, which acted as a cathode. Figure 4 shows that the corroding bar (bar number 4) was anodically polarized and its potential increased from -800 mV/SCE to -660 mV/SCE, while the potential of passive bars in saturated concrete and in dry concrete decreased from its free corrosion value, respectively, of 240 mV and 400 mV. Figure 5 shows the current density on the surface of steel bars in saturated concrete (triangles) and in dry concrete (circles). Passive steel bars had a cathodic current density (white symbols), and because of the higher oxygen content of dry concrete, bars in this zone received a cathodic current density one order of magnitude higher than in saturated concrete.

Meanwhile, the depassivated steel bar (bar number 4) received an anodic current density of about 15 mA/m² (black symbol). Considering that in the absence of the macrocell the corrosion rate (i_{CORR}) of this bar embedded in water saturated concrete was negligible, the macrocell current density (i_{MACRO}) equals to the corrosion rate in the presence of macrocell.

Another effect of the onset of macrocell is the decrease of the potential of passive steel bars in saturated concrete: in fact it decreased by 240 mV from -400 mV/SCE to -640 mV/SCE (Figure 4). Due to this cathodic polarization, these bars benefit of cathodic prevention effect, which increase the CTL. Hence, this simulation showed

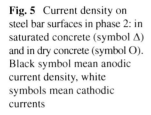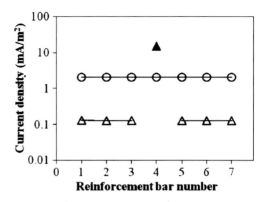

Fig. 5 Current density on steel bar surfaces in phase 2: in saturated concrete (symbol Δ) and in dry concrete (symbol O). Black symbol mean anodic current density, white symbols mean cathodic currents

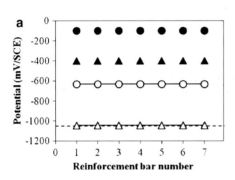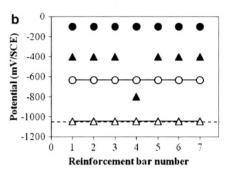

Fig. 6 Potential of protected steel bars, respectively, in phase 1 (**a**) and in phase 2 (**b**): in saturated concrete (symbol Δ) and in dry concrete (symbol O) in free corrosion conditions (black symbols) and protected with a sacrificial anode (white symbols). The dashed line shows the anode potential

that once corrosion initiates a sharp increase in active steel corrosion rate takes place, but, in the meantime, corrosion of the active bar may protect passive steel areas.

3.2 Wall with sacrificial anode

Due to the protection effect shown by a corroding bar on the passive bars, simulations were also carried out to study the effectiveness of cathodic prevention and cathodic protection applied by means of sacrificial anodes placed in seawater. Results showed that sacrificial anodes may be effective both in preventing steel depassivation (cathodic prevention) or controlling steel corrosion rate after its depassivation (cathodic protection). Figure 6 shows the results obtained applying a sacrificial anode to the structure before steel depassivation (phase 1) and after steel depassivation (phase 2). This figure shows the potential of steel bars in dry concrete in free corrosion condition (black circles) and when protected with the anode (white circles). It also shows the potential of steel bars in saturated concrete in free corrosion condition (black triangles) and when protected with the anode (white triangles): the difference between these values is the cathodic polarization induced by the external sacrificial anode. The dashed line shows the external zinc anode potential (E_{ANODE} = -1050 mV/SCE). Because of the low electrical resistivity of seawater, the potential of all steel bars in saturated concrete approaches the anode potential (about -1040 mV/SCE) irrespective of passive or active condition of steel. For this reason, the results obtained with a cathodic prevention system applied before steel depassivation (Figure 6a) are similar to the results obtained with a cathodic protection system (Figure 6b). Consequently, in the peculiar situation of hollow structures, the effectiveness of a cathodic prevention system and a cathodic protection system could be compared. Indeed, both in phase 1 and in phase 2, steel is brought to a perfect passivity condition, where pitting corrosion could not initiate nor propagate [1]. However, to discuss when a cathodic system should be applied to hollow structures, also other aspects have to be considered. Indeed, a cathodic prevention system could avoid macrocell development and so it could avoid steel depassivation. It has to be considered that it is a risk to install a protection system after steel depassivation, because of its high corrosion rate: in this case it would be necessary to install a corrosion monitoring system and to provide a cathodic protection system immediately when steel depassivation occurs. In the end, it is certainly easier to install sacrificial anodes during the construction phase.

Nevertheless, it has to be highlighted that the consumption rate of sacrificial anodes is higher on hollow structures than on completely saturated structures. Indeed in the modelling the consumption rate of the anode in contact with seawater exceeded 15 mA/m^2 even in the absence of corroding bars. This high depletion rate of the anode was due to the current fed by the anode which was not confined to the external reinforcement layer, but it reached also steel bars in dry concrete, which underwent a high cathodic polarization (Figure 6). These bars have in practice no cathodic limiting current density, because of the high oxygen content of dry concrete, so to polarize these bars the anode had to supply a current about 30 times higher than that required to polarize the bars in saturated concrete.

4 Conclusions and future developments

Results showed that, in submerged hollow structures, macrocells can both promote steel depassivation during the initiation period and increase corrosion rate on active areas after steel depassivation. However, the effects of macrocells could be controlled by installing cathodic prevention or cathodic protection. Indeed, steel depassivation as well as corrosion rate of active area can be inhibited by coupling reinforcement bars with sacrificial anodes placed in seawater. However, these results suggested new questions or open issues that need further studies.

1. In the models described in this paper, the macrocell among one depassivated steel bar and 13 still passive bars was studied, so the anode-to-cathode area ratio

was imposed. What could be a realistic ratio in this structure has not been studied yet. However, a critical value for this parameter could exist: indeed the macrocell current (I_{MACRO}) supplied by the active anodic area has to ensure the cathodic polarization of still passive steel bars to prevent their depassivation. If the active area is not wide enough, the protection of passive steel cannot be ensured and the anodic area has to increase.

2. Corrosion and protection of steel reinforcement in portions of structures emerging from seawater was not studied yet. Generally, in the splash zone of structures steel depassivation is promoted because of the simultaneous presence of chlorides and oxygen. However, in hollow structures steel depassivation could start in the submerged zone, so it has to be investigated how steel reinforcement corrosion could initiate and propagate in this particular type of structures. Also if a sacrificial anode placed in seawater could protect steel reinforcement in the emerged portion of a hollow structure was not studied yet.

3. Results have shown that in hollow structures the consumption rate of sacrificial anodes could be a relevant problem, because most of the supplied current is wasted to polarize passive steel bars in dry concrete. For this reason, the real behaviour of sacrificial anodes in time needs to be investigated further and the efficiency of other systems could be studied, for instance taking into account cathodic prevention or cathodic protection systems with impressed current.

4. One of the main parameters affecting the results is the humidity profile in the concrete wall: indeed, it affects both the limiting current density due to oxygen diffusion and the electrical resistivity of concrete. For this reason, it would be necessary to understand the effects of the parameters affecting the humidity profile in hollow structures: indeed in these structures a particular mechanism of water transport could exist, because of the water pressure on the external side of the element, the unsaturated water flow in concrete and the water evaporation on the internal side of the element. It should be considered also that in these structures chlorides penetration could be promoted by capillary suction of seawater.

5. Finally, in order to evaluate the real risks of macrocell and advantages of sacrificial anodes applied in the initiation stage, the dependence of the chloride threshold on the steel potential (and thus on the exposure condition of the concrete which determines its moisture content) has to be investigated. In this case, however, the evaluation of the chloride threshold is even more difficult than under the usual conditions of exposure to the atmosphere. In fact, the high value expected for the CTL in water saturated concrete makes it in practice unfeasible to use testing methods based on the penetration of chlorides through the concrete.

References

[1] Pedeferri P., (1996), *Constr. Build. Mater.*, vol. 10, n. 5, pp. 391–402.
[2] Alonso C., Castellote M., Andrade C., (2002), *Electrochim. Acta*, vol. 47, n. 21, pp. 3469–3481.

[3] Haugen T., Berge H. E., Espelid B., Markey I., (2009), Proceedings of the International Conference Eurocorr, pp. 1–7, Nice.
[4] Raupach M., Polder R., Frolund T., Nygaard P., (2007), Proceedings of the International Conference Eurocorr, pp. 1–9, Freiburg.
[5] Warkus J., Raupach M., (2006), *Mater. Corros.*, vol. 57, n. 12, pp. 920–925.
[6] Warkus J., Raupach M., (2008), *Mater. Corros.*, vol. 59, n. 2, pp. 122–130.
[7] Bertolini L., Redaelli E., (2009), *Corros. Sci.*, vol. 51, n. 9, pp. 2218–2230.
[8] Bertolini L., Redaelli E., (2004), Proceedings of the International Conference Eurocorr, pp. 1–10, Nice.
[9] Redaelli E., Bertolini L., Peelen W., Polder R., (2006), *Mater. Corros.*, vol. 57, n. 8, pp. 628–635.
[10] Kranc S.C., Sagüés A.A., (2001), *Mater. Corros.*, vol. 43, n. 7, pp. 1355–1372.
[11] Gulikers J., Raupach M., (2006), *Mater. Corros.*, vol. 57, n. 8, pp. 618–627.
[12] Muehlenkamp, E. B., Koretsky, M. D., Westall, J.C., (2005), *Corrosion*, vol. 61, n. 6, pp. 519–533.

RILEM Publications

The following list is presenting our global offer, sorted by series.

RILEM PROCEEDINGS

PRO 1: Durability of High Performance Concrete (1994) 266 pp., ISBN: 2-91214-303-9; e-ISBN: 2-35158-012-5; *Ed. H. Sommer*

PRO 2: Chloride Penetration into Concrete (1995) 496 pp., ISBN: 2-912143-00-4; e-ISBN: 2-912143-45-4; *Eds. L.-O. Nilsson and J.-P. Ollivier*

PRO 3: Evaluation and Strengthening of Existing Masonry Structures (1995) 234 pp., ISBN: 2-912143-02-0; e-ISBN: 2-351580-14-1; *Eds. L. Binda and C. Modena*

PRO 4: Concrete: From Material to Structure (1996) 360 pp., ISBN: 2-912143-04-7; e-ISBN: 2-35158-020-6; *Eds. J.-P. Bournazel and Y. Malier*

PRO 5: The Role of Admixtures in High Performance Concrete (1999) 520 pp., ISBN: 2-912143-05-5; e-ISBN: 2-35158-021-4; *Eds. J. G. Cabrera and R. Rivera-Villarreal*

PRO 6: High Performance Fiber Reinforced Cement Composites (HPFRCC 3) (1999) 686 pp., ISBN: 2-912143-06-3; e-ISBN: 2-35158-022-2; *Eds. H. W. Reinhardt and A. E. Naaman*

PRO 7: 1st International RILEM Symposium on Self-Compacting Concrete (1999) 804 pp., ISBN: 2-912143-09-8; e-ISBN: 2-912143-72-1; *Eds. Å. Skarendahl and Ö. Petersson*

PRO 8: International RILEM Symposium on Timber Engineering (1999) 860 pp., ISBN: 2-912143-10-1; e-ISBN: 2-35158-023-0; *Ed. L. Boström*

PRO 9: 2nd International RILEM Symposium on Adhesion between Polymers and Concrete ISAP '99 (1999) 600 pp., ISBN: 2-912143-11-X; e-ISBN: 2-35158-024-9; *Eds. Y. Ohama and M. Puterman*

PRO 10: 3rd International RILEM Symposium on Durability of Building and Construction Sealants (2000) 360 pp., ISBN: 2-912143-13-6; e-ISBN: 2-351580-25-7; *Eds. A. T. Wolf*

PRO 11: 4th International RILEM Conference on Reflective Cracking in Pavements (2000) 549 pp., ISBN: 2-912143-14-4; e-ISBN: 2-35158-026-5; *Eds. A. O. Abd El Halim, D. A. Taylor and El H. H. Mohamed*

PRO 12: International RILEM Workshop on Historic Mortars: Characteristics and Tests (1999) 460 pp., ISBN: 2-912143-15-2; e-ISBN: 2-351580-27-3; *Eds. P. Bartos, C. Groot and J. J. Hughes*

PRO 13: 2nd International RILEM Symposium on Hydration and Setting (1997) 438 pp., ISBN: 2-912143-16-0; e-ISBN: 2-35158-028-1; *Ed. A. Nonat*

PRO 14: Integrated Life-Cycle Design of Materials and Structures (ILCDES 2000) (2000) 550 pp., ISBN: 951-758-408-3; e-ISBN: 2-351580-29-X, ISSN: 0356–9403; *Ed. S. Sarja*

PRO 15: Fifth RILEM Symposium on Fibre-Reinforced Concretes (FRC) – BEFIB'2000 (2000) 810 pp., ISBN: 2-912143-18-7; e-ISBN: 2-912143-73-X; *Eds. P. Rossi and G. Chanvillard*

PRO 16: Life Prediction and Management of Concrete Structures (2000) 242 pp., ISBN: 2-912143-19-5; e-ISBN: 2-351580-30-3; *Ed. D. Naus*

PRO 17: Shrinkage of Concrete – Shrinkage 2000 (2000) 586 pp., ISBN: 2-912143-20-9; e-ISBN: 2-351580-31-1; *Eds. V. Baroghel-Bouny and P.-C. Aïtcin*

PRO 18: Measurement and Interpretation of the On-Site Corrosion Rate (1999) 238 pp., ISBN: 2-912143-21-7; e-ISBN: 2-351580-32-X; *Eds. C. Andrade, C. Alonso, J. Fullea, J. Polimon and J. Rodriguez*

PRO 19: Testing and Modelling the Chloride Ingress into Concrete (2000) 516 pp., ISBN: 2-912143-22-5; e-ISBN: 2-351580-33-8; Soft cover, *Eds. C. Andrade and J. Kropp*

PRO 20: 1st International RILEM Workshop on Microbial Impacts on Building Materials (2000) 74 pp., e-ISBN: 2-35158-013-3; *Ed. M. Ribas Silva (CD 02)*

PRO 21: International RILEM Symposium on Connections between Steel and Concrete (2001) 1448 pp., ISBN: 2-912143-25-X; e-ISBN: 2-351580-34-6; *Ed. R. Eligehausen*

PRO 22: International RILEM Symposium on Joints in Timber Structures (2001) 672 pp., ISBN: 2-912143-28-4; e-ISBN: 2-351580-35-4; *Eds. S. Aicher and H.-W. Reinhardt*

PRO 23: International RILEM Conference on Early Age Cracking in Cementitious Systems (2003) 398 pp., ISBN: 2-912143-29-2; e-ISBN: 2-351580-36-2; *Eds. K. Kovler and A. Bentur*

PRO 24: 2nd International RILEM Workshop on Frost Resistance of Concrete (2002) 400 pp., ISBN: 2-912143-30-6; e-ISBN: 2-351580-37-0, Hard back; *Eds. M. J. Setzer, R. Auberg and H.-J. Keck*

PRO 25: International RILEM Workshop on Frost Damage in Concrete (1999) 312 pp., ISBN: 2-912143-31-4; e-ISBN: 2-351580-38-9, Soft cover; *Eds. D. J. Janssen, M. J. Setzer and M. B. Snyder*

PRO 26: International RILEM Workshop on On-Site Control and Evaluation of Masonry Structures (2003) 386 pp., ISBN: 2-912143-34-9; e-ISBN: 2-351580-14-1, Soft cover; *Eds. L. Binda and R. C. de Vekey*

PRO 27: International RILEM Symposium on Building Joint Sealants (1988) 240 pp., e-ISBN: 2-351580-15-X; *Ed. A. T. Wolf, (CD03)*

PRO 28: 6th International RILEM Symposium on Performance Testing and Evaluation of Bituminous Materials, PTEBM'03, Zurich, Switzerland (2003) 652 pp., ISBN: 2-912143-35-7; e-ISBN: 2-912143-77-2, Soft cover; *Ed. M. N. Partl (CD06)*

PRO 29: 2nd International RILEM Workshop on Life Prediction and Ageing Management of Concrete Structures, Paris, France (2003) 402 pp., ISBN: 2-912143-36-5; e-ISBN: 2-912143-78-0, Soft cover; *Ed. D. J. Naus*

PRO 30: 4th International RILEM Workshop on High Performance Fiber Reinforced Cement Composites – HPFRCC 4, University of Michigan, Ann Arbor, USA (2003) 562 pp., ISBN: 2-912143-37-3; e-ISBN: 2-912143-79-9, Hard back; *Eds. A. E. Naaman and H. W. Reinhardt*

PRO 31: International RILEM Workshop on Test and Design Methods for Steel Fibre Reinforced Concrete: Background and Experiences (2003) 230 pp., ISBN: 2-912143-38-1; e-ISBN: 2-351580-16-8, Soft cover; *Eds. B. Schnütgen and L. Vandewalle*

PRO 32: International Conference on Advances in Concrete and Structures, 2 volumes (2003) 1592 pp., ISBN (set): 2-912143-41-1; e-ISBN: 2-351580-17-6, Soft cover; *Eds. Ying-shu Yuan, Surendra P. Shah and Heng-lin Lü*

PRO 33: 3rd International Symposium on Self-Compacting Concrete (2003) 1048 pp., ISBN: 2-912143-42-X; e-ISBN: 2-912143-71-3, Soft cover; *Eds. Ó. Wallevik and I. Níelsson*

PRO 34: International RILEM Conference on Microbial Impact on Building Materials (2003) 108 pp., ISBN: 2-912143-43-8; e-ISBN: 2-351580-18-4; *Ed. M. Ribas Silva*

PRO 35: International RILEM TC 186-ISA on Internal Sulfate Attack and Delayed Ettringite Formation (2002) 316 pp., ISBN: 2-912143-44-6; e-ISBN: 2-912143-80-2, Soft cover; *Eds. K. Scrivener and J. Skalny*

PRO 36: International RILEM Symposium on Concrete Science and Engineering – A Tribute to Arnon Bentur (2004) 264 pp., ISBN: 2-912143-46-2; e-ISBN: 2-912143-58-6, Hard back; *Eds. K. Kovler, J. Marchand, S. Mindess and J. Weiss*

PRO 37: 5th International RILEM Conference on Cracking in Pavements – Mitigation, Risk Assessment and Prevention (2004) 740 pp., ISBN: 2-912143-47-0; e-ISBN: 2-912143-76-4, Hard back; *Eds. C. Petit, I. Al-Qadi and A. Millien*

PRO 38: 3rd International RILEM Workshop on Testing and Modelling the Chloride Ingress into Concrete (2002) 462 pp., ISBN: 2-912143-48-9; e-ISBN: 2-912143-57-8, Soft cover; *Eds. C. Andrade and J. Kropp*

PRO 39: 6th International RILEM Symposium on Fibre-Reinforced Concretes (BEFIB 2004), 2 volumes, (2004) 1536 pp., ISBN: 2-912143-51-9 (set); e-ISBN: 2-912143-74-8, Hard back; *Eds. M. Di Prisco, R. Felicetti and G. A. Plizzari*

PRO 40: International RILEM Conference on the Use of Recycled Materials in Buildings and Structures (2004) 1154 pp., ISBN: 2-912143-52-7 (set); e-ISBN: 2-912143-75-6, Soft cover; *Eds. E. Vázquez, Ch. F. Hendriks and G. M. T. Janssen*

PRO 41: RILEM International Symposium on Environment-Conscious Materials and Systems for Sustainable Development (2005) 450 pp., ISBN: 2-912143-55-1; e-ISBN: 2-912143-64-0, Soft cover; *Eds. N. Kashino and Y. Ohama*

PRO 42: SCC'2005 – China: 1st International Symposium on Design, Performance and Use of Self-Consolidating Concrete (2005) 726 pp., ISBN: 2-912143-61-6; e-ISBN: 2-912143-62-4, Hard back; *Eds. Zhiwu Yu, Caijun Shi, Kamal Henri Khayat and Youjun Xie*

PRO 43: International RILEM Workshop on Bonded Concrete Overlays (2004) 114 pp., e-ISBN: 2-912143-83-7; *Eds. J. L. Granju and J. Silfwerbrand*

PRO 44: 2nd International RILEM Workshop on Microbial Impacts on Building Materials (Brazil 2004) (CD11) 90 pp., e-ISBN: 2-912143-84-5; *Ed. M. Ribas Silva*

PRO 45: 2nd International Symposium on Nanotechnology in Construction, Bilbao, Spain (2005) 414 pp., ISBN: 2-912143-87-X; e-ISBN: 2-912143-88-8, Soft cover; *Eds. Peter J. M. Bartos, Yolanda de Miguel and Antonio Porro*

PRO 46: ConcreteLife'06 – International RILEM-JCI Seminar on Concrete Durability and Service Life Planning: Curing, Crack Control, Performance in Harsh Environments (2006) 526 pp., ISBN: 2-912143-89-6; e-ISBN: 2-912143-90-X, Hard back; *Ed. K. Kovler*

PRO 47: International RILEM Workshop on Performance Based Evaluation and Indicators for Concrete Durability (2007) 385 pp., ISBN: 978-2-912143-95-2; e-ISBN: 978-2-912143-96-9, Soft cover; *Eds. V. Baroghel-Bouny, C. Andrade, R. Torrent and K. Scrivener*

PRO 48: 1st International RILEM Symposium on Advances in Concrete through Science and Engineering (2004) 1616 pp., e-ISBN: 2-912143-92-6; *Eds. J. Weiss, K. Kovler, J. Marchand, and S. Mindess*

PRO 49: International RILEM Workshop on High Performance Fiber Reinforced Cementitious Composites in Structural Applications (2006) 598 pp., ISBN: 2-912143-93-4; e-ISBN: 2-912143-94-2, Soft cover; *Eds. G. Fischer and V.C. Li*

PRO 50: 1st International RILEM Symposium on Textile Reinforced Concrete (2006) 418 pp., ISBN: 2-912143-97-7; e-ISBN: 2-351580-08-7, Soft cover; *Eds. Josef Hegger, Wolfgang Brameshuber and Norbert Will*

PRO 51: 2nd International Symposium on Advances in Concrete through Science and Engineering (2006) 462 pp., ISBN: 2-35158-003-6; e-ISBN: 2-35158-002-8, Hard back; *Eds. J. Marchand, B. Bissonnette, R. Gagné, M. Jolin and F. Paradis*

PRO 52: Volume Changes of Hardening Concrete: Testing and Mitigation (2006) 428 pp., ISBN: 2-35158-004-4; e-ISBN: 2-35158-005-2, Soft cover; *Eds. O. M. Jensen, P. Lura and K. Kovler*

PRO 53: High Performance Fiber Reinforced Cement Composites HPFRCC5 (2007) 542 pp., ISBN: 978-2-35158-046-2; e-ISBN: 978-2-35158-089-9, Hard back; *Eds. H. W. Reinhardt and A. E. Naaman*

PRO 54: 5th International RILEM Symposium on Self-Compacting Concrete, 3 Volumes (2007) 1198 pp., ISBN: 978-2-35158-047-9; e-ISBN: 978-2-35158-088-2, Soft cover; *Eds. G. De Schutter and V. Boel*

PRO 55: International RILEM Symposium Photocatalysis, Environment and Construction Materials (2007) 350 pp., ISBN: 978-2-35158-056-1; e-ISBN: 978-2-35158-057-8, Soft cover; *Eds. P. Baglioni and L. Cassar*

PRO 56: International RILEM Workshop on Integral Service Life Modelling of Concrete Structures (2007) 458 pp., ISBN 978-2-35158-058-5; e-ISBN: 978-2-35158-090-5, Hard back; *Eds. R. M. Ferreira, J. Gulikers and C. Andrade*

PRO 57: RILEM Workshop on Performance of cement-based materials in aggressive aqueous environments (2008) 132 pp., e-ISBN: 978-2-35158-059-2; *Ed. N. De Belie*

PRO 58: International RILEM Symposium on Concrete Modelling CONMOD'08 (2008) 847 pp., ISBN: 978-2-35158-060-8; e-ISBN: 978-2-35158-076-9, Soft cover; *Eds. E. Schlangen and G. De Schutter*

PRO 59: International RILEM Conference on On Site Assessment of Concrete, Masonry and Timber Structures SACoMaTiS 2008, 2 volumes (2008) 1232 pp., ISBN: 978-2-35158-061-5 (set); e-ISBN: 978-2-35158-075-2, Hard back; *Eds. L. Binda, M. di Prisco and R. Felicetti*

PRO 60: Seventh RILEM International Symposium (BEFIB 2008) on Fibre Reinforced Concrete: Design and Applications (2008) 1181 pp, ISBN: 978-2-35158-064-6; e-ISBN: 978-2-35158-086-8, Hard back; *Ed. R. Gettu*

PRO 61: 1st International Conference on Microstructure Related Durability of Cementitious Composites (Nanjing), 2 volumes, (2008) 1524 pp., ISBN: 978-2-35158-065-3; e-ISBN: 978-2-35158-084-4; *Eds. W. Sun, K. van Breugel, C. Miao, G. Ye and H. Chen*

PRO 62: NSF/ RILEM Workshop: In-situ Evaluation of Historic Wood and Masonry Structures (2008) 130 pp., e-ISBN: 978-2-35158-068-4; *Eds. B. Kasal, R. Anthony and M. Drdácký*

PRO 63: Concrete in Aggressive Aqueous Environments: Performance, Testing and Modelling, 2 volumes, (2009) 631 pp., ISBN: 978-2-35158-071-4; e-ISBN: 978-2-35158-082-0, Soft cover; *Eds. M. G. Alexander and A. Bertron*

PRO 64: Long Term Performance of Cementitious Barriers and Reinforced Concrete in Nuclear Power Plants and Waste Management – NUCPERF 2009 (2009) 359 pp., ISBN: 978-2-35158-072-1; e-ISBN: 978-2-35158-087-5; *Eds. V. L'Hostis, R. Gens, C. Gallé*

PRO 65: Design Performance and Use of Self-consolidating Concrete, SCC'2009, (2009) 913 pp., ISBN: 978-2-35158-073-8; e-ISBN: 978-2-35158-093-6; *Eds. C. Shi, Z. Yu, K. H. Khayat and P. Yan*

PRO 66: Concrete Durability and Service Life Planning, 2nd International RILEM Workshop, ConcreteLife'09, (2009) 626 pp., ISBN: 978-2-35158-074-5; e-ISBN: 978-2-35158-085-1; *Ed. K. Kovler*

PRO 67: Repairs Mortars for Historic Masonry (2009) 397 pp., e-ISBN: 978-2-35158-083-7; *Ed. C. Groot*

PRO 68: Proceedings of the 3rd International RILEM Symposium on 'Rheology of Cement Suspensions such as Fresh Concrete' (2009) 372 pp., ISBN: 978-2-35158-091-2; e-ISBN: 978-2-35158-092-9; *Eds. O. H. Wallevik, S. Kubens and S. Oesterheld*

PRO 69: 3rd International PhD Student Workshop on 'Modelling the Durability of Reinforced Concrete' (2009) 122 pp., ISBN: 978-2-35158-095-0; e-ISBN: 978-2-35158-094-3; *Eds. R. M. Ferreira, J. Gulikers and C. Andrade*

PRO 71: Advances in Civil Engineering Materials, Proceedings of the 'The 50-year Teaching Anniversary of Prof. Sun Wei', (2010) 307 pp., ISBN: 978-2-35158-098-1; e-ISBN: 978-2-35158-099-8; *Eds. C. Miao, G. Ye, and H. Chen*

PRO 74: International RILEM Conference on 'Use of Superabsorsorbent Polymers and Other New Additives in Concrete' (2010) 374 pp., ISBN: 978-2-35158-104-9; e-ISBN: 978-2-35158-105-6; *Eds. O.M. Jensen, M.T. Hasholt, and S. Laustsen*

PRO 75: International Conference on 'Material Science - 2nd ICTRC - Textile Reinforced Concrete - Theme 1' (2010) 436 pp., ISBN: 978-2-35158-106-3; e-ISBN: 978-2-35158-107-0; *Ed. W. Brameshuber*

PRO 76: International Conference on 'Material Science - HetMat - Modelling of Heterogeneous Materials - Theme 2' (2010) 255 pp., ISBN: 978-2-35158-108-7; e-ISBN: 978-2-35158-109-4; *Ed. W. Brameshuber*

PRO 77: International Conference on 'Material Science - AdIPoC - Additions Improving Properties of Concrete - Theme 3' (2010) 459 pp., ISBN: 978-2-35158-110-0; e-ISBN: 978-2-35158-111-7; *Ed. W. Brameshuber*

PRO 78: 2nd Historic Mortars Conference and RILEM TC 203-RHM Final Workshop – HMC2010 (2010) 1416 pp., e-ISBN: 978-2-35158-112-4; *Eds J. Válek, C. Groot, and J. J. Hughes*

PRO 79: International RILEM Conference on Advances in Construction Materials Through Science and Engineering (2011) 213 pp., e-ISBN: 978-2-35158-117-9; *Eds Christopher Leung and K.T. Wan*

PRO 80: 2nd International RILEM Conference on Concrete Spalling due to Fire Exposure (2011) 453 pp., ISBN: 978-2-35158-118-6, e-ISBN: 978-2-35158-119-3; *Eds E.A.B. Koenders and F. Dehn*

RILEM REPORTS

Report 19: Considerations for Use in Managing the Aging of Nuclear Power Plant Concrete Structures (1999) 224 pp., ISBN: 2-912143-07-1; e-ISBN: 2-35158-039-7; *Ed. D. J. Naus*

Report 20: Engineering and Transport Properties of the Interfacial Transition Zone in Cementitious Composites (1999) 396 pp., ISBN: 2-912143-08-X; e-ISBN: 2-35158-040-0; *Eds. M. G. Alexander, G. Arliguie, G. Ballivy, A. Bentur and J. Marchand*

Report 21: Durability of Building Sealants (1999) 450 pp., ISBN: 2-912143-12-8; e-ISBN: 2-35158-041-9; *Ed. A. T. Wolf*

Report 22: Sustainable Raw Materials – Construction and Demolition Waste (2000) 202 pp., ISBN: 2-912143-17-9; e-ISBN: 2-35158-042-7; *Eds. C. F. Hendriks and H. S. Pietersen*

Report 23: Self-Compacting Concrete state-of-the-art report (2001) 166 pp., ISBN: 2-912143-23-3; e-ISBN: 2-912143-59-4, Soft cover; *Eds. Å. Skarendahl and Ö. Petersson*

Report 24: Workability and Rheology of Fresh Concrete: Compendium of Tests (2002) 154 pp., ISBN: 2-912143-32-2; e-ISBN: 2-35158-043-5, Soft cover; *Eds. P. J. M. Bartos, M. Sonebi and A. K. Tamimi*

Report 25: Early Age Cracking in Cementitious Systems (2003) 350 pp., ISBN: 2-912143-33-0; e-ISBN: 2-912143-63-2, Soft cover; *Ed. A. Bentur*

Report 26: Towards Sustainable Roofing (Joint Committee CIB/RILEM) (CD 07), (2001) 28 pp., e-ISBN: 2-912143-65-9; *Eds. Thomas W. Hutchinson and Keith Roberts*

Report 27: Condition Assessment of Roofs (Joint Committee CIB/RILEM) (CD 08), (2003) 12 pp., e-ISBN: 2-912143-66-7

Report 28: Final report of RILEM TC 167-COM 'Characterisation of Old Mortars with Respect to Their Repair' (2007) 192 pp., ISBN: 978-2-912143-56-3; e-ISBN: 978-2-912143-67-9, Soft cover; *Eds. C. Groot, G. Ashall and J. Hughes*

Report 29: Pavement Performance Prediction and Evaluation (PPPE): Interlaboratory Tests (2005) 194 pp., e-ISBN: 2-912143-68-3; *Eds. M. Partl and H. Piber*

Report 30: Final Report of RILEM TC 198-URM 'Use of Recycled Materials' (2005) 74 pp., ISBN: 2-912143-82-9; e-ISBN: 2-912143-69-1 – Soft cover; *Eds. Ch. F. Hendriks, G. M. T. Janssen and E. Vázquez*

Report 31: Final Report of RILEM TC 185-ATC 'Advanced testing of cement-based materials during setting and hardening' (2005) 362 pp., ISBN: 2-912143-81-0; e-ISBN: 2-912143-70-5 – Soft cover; *Eds. H. W. Reinhardt and C. U. Grosse*

Report 32: Probabilistic Assessment of Existing Structures. A JCSS publication (2001) 176 pp., ISBN 2-912143-24-1; e-ISBN: 2-912143-60-8 – Hard back; *Ed. D. Diamantidis*

Report 33: State-of-the-Art Report of RILEM Technical Committee TC 184-IFE 'Industrial Floors' (2006) 158 pp., ISBN 2-35158-006-0; e-ISBN: 2-35158-007-9, Soft cover; *Ed. P. Seidler*

Report 34: Report of RILEM Technical Committee TC 147-FMB 'Fracture mechanics applications to anchorage and bond' Tension of Reinforced Concrete Prisms – Round Robin Analysis and Tests on Bond (2001) 248 pp., e-ISBN 2-912143-91-8; *Eds. L. Elfgren and K. Noghabai*

Report 35: Final Report of RILEM Technical Committee TC 188-CSC 'Casting of Self Compacting Concrete' (2006) 40 pp., ISBN 2-35158-001-X; e-ISBN: 2-912143-98-5 – Soft cover;*Eds. Å. Skarendahl and P. Billberg*

Report 36: State-of-the-Art Report of RILEM Technical Committee TC 201-TRC 'Textile Reinforced Concrete' (2006) 292 pp., ISBN 2-912143-99-3; e-ISBN: 2-35158-000-1, Soft cover; *Ed. W. Brameshuber*

Report 37: State-of-the-Art Report of RILEM Technical Committee TC 192-ECM 'Environment-conscious construction materials and systems' (2007) 88 pp., ISBN: 978-2-35158-053-0; e-ISBN: 2-35158-079-0, Soft cover; *Eds. N. Kashino, D. Van Gemert and K. Imamoto*

Report 38: State-of-the-Art Report of RILEM Technical Committee TC 205-DSC 'Durability of Self-Compacting Concrete' (2007) 204 pp., ISBN: 978-2-35158-048-6; e-ISBN: 2-35158-077-6, Soft cover; *Eds. G. De Schutter and K. Audenaert*

Report 39: Final Report of RILEM Technical Committee TC 187-SOC 'Experimental determination of the stress-crack opening curve for concrete in tension' (2007) 54 pp., ISBN 978-2-35158-049-3; e-ISBN: 978-2-35158-078-3, Soft cover; *Ed. J. Planas*

Report 40: State-of-the-Art Report of RILEM Technical Committee TC 189-NEC 'Non-Destructive Evaluation of the Penetrability and Thickness of the Concrete Cover' (2007) 246 pp., ISBN 978-2-35158-054-7; e-ISBN: 978-2-35158-080-6, Soft cover; *Eds. R. Torrent and L. Fernández Luco*

Report 41: State-of-the-Art Report of RILEM Technical Committee TC 196-ICC 'Internal Curing of Concrete' (2007) 164 pp., ISBN: 978-2-35158-009-7; e-ISBN: 978-2-35158-082-0, Soft cover; *Eds. K. Kovler and O. M. Jensen*

Report 42: 'Acoustic Emission and Related Non-destructive Evaluation Techniques for Crack Detection and Damage Evaluation in Concrete' – Final Report of RILEM Technical Committee 212-ACD (2010) 12 pp., e-ISBN: 978-2-35158-100-1; *Ed. M. Ohtsu*

RILEM Publications published by Springer

RILEM BOOKSERIES (Proceedings)

VOL. 1: Design, Production and Placement of Self-Consolidating Concrete (2010) 466 pp., ISBN: 978-90-481-9663-0; e-ISBN: 978-90-481-9664-7, Hardcover; *Ed. K. Khayat and D. Feyes*

VOL. 5: Joint fib-RILEM Workshop on Modelling of Corroding Concrete Structures (2011) 290 pp., ISBN: 978-94-007-0676-7; e-ISBN: 978-94-007-0677-4, Hardcover; *Ed. C. Andrade and G. Mancini*

For the latest publications in the RILEM Bookseries, please visit http://www.springer.com/series/8781

RILEM STATE-OF-THE-ART REPORTS

VOL. 3: State-of-the-Art Report of RILEM Technical Committee TC 193-RLS 'Bonded Cement-Based Material Overlays for the Repair, the Lining or the Strengthening of Slabs or Pavements' (2011) 198 pp., ISBN: 978-94-007-1238-6; e-ISBN: 978-94-007-1239-3, Hardcover; *Ed. B. Bissonnette, L. Courard, D.W. Fowler and J-L. Granju*

VOL. 4: State-of-the-Art Report prepared by Subcommittee 2 of RILEM Technical Committee TC 208-HFC 'Durability of Strain-Hardening Fibre-Reinforced Cement-Based Composites' (SHCC) (2011) 151 pp., ISBN: 978-94-007-0337-7; e-ISBN: 978-94-007-0338-4, Hardcover; *Ed. G.P.A.G. van Zijl and F.H. Wittmann*

VOL. 5: State-of-the-Art Report of RILEM Technical Committee TC 194-TDP 'Application of Titanium Dioxide Photocatalysis to Construction Materials' (2011) 60 pp., ISBN: 978-94-007-1296-6; e-ISBN: 978-94-007-1297-3, Hardcover; *Ed. Yoshihiko Ohama and Dionys Van Gemert*

VOL. 7: State-of-the-Art Report of RILEM Technical Committee TC 215-AST 'In Situ Assessment of Structural Timber' (2010) 152 pp., ISBN: 978-94-007-0559-3; e-ISBN: 978-94-007-0560-9, Hardcover; *Ed. B. Kasal and T. Tannert*

For the latest publications in the RILEM State-of-the-Art Reports, please visit http://www.springer.com/series/8780